THE ULTIMATE GUIDE TO POWER TOOLS

MASTER EFFICIENCY, ENSURE SAFETY & PERFECT MAINTENANCE

UNLOCK THE FULL POTENTIAL, CREATIVITY AND PRODUCTIVITY WITH EXPERT INSIGHTS AND TIPS

KUNAL GARG
MAYANK GARG

INDIA • SINGAPORE • MALAYSIA

ISBN 979-8-89519-838-4

This book is dedicated to our Father, ***Late Dr. G.P.Garg.***

Dr. G.P. Garg began his career in hand tools in 1980 in Old Delhi. His vision and hard work laid the foundation for our family's success. In 2006, our father ventured into the power tools business and entrusted it to us, enabling us to continue his legacy. His entrepreneurial spirit and dedication continue to inspire us every day, driving us to innovate and serve the needs of carpenters, electricians, plumbers, metalworkers and construction workers.

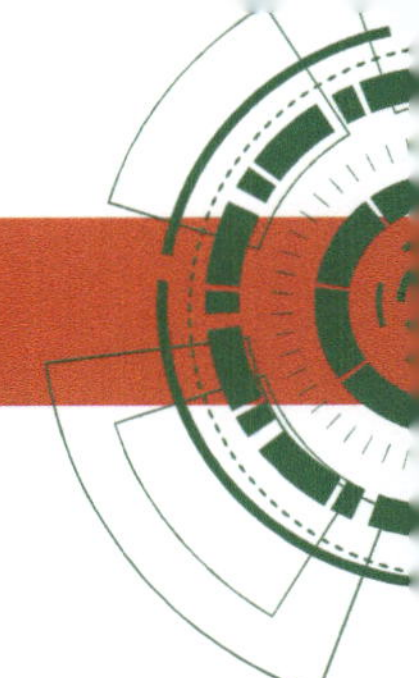

Acknowledgements

We extend our heartfelt gratitude to our partner, **Mr. Anand Kumar Jain.** His extensive life experience, wisdom, and business acumen have been invaluable in guiding us through the complexities of the power tools industry. His support and mentorship have been critical to our success, and we are deeply thankful for his unwavering belief in us. Your dedication to excellence inspires us to continually strive for better solutions for the construction industry.

Foreword

The world of power tools is vast and ever-evolving, offering endless possibilities for those who seek to harness their power and precision. As someone who has spent decades in the industry, I have seen how the right tools can transform not only the quality of work but also the efficiency and safety of the user.

Kunal and Mayank Garg bring deep expertise to this comprehensive guide. Their hands-on experience and extensive research make this an invaluable resource for anyone looking to master power tools. From practical advice to expert tips, this book provides everything needed to achieve excellence in your projects.

I'm honoured to introduce this guide and confident that it will inspire and equip you to unlock the full potential of power tools.

Mr Anand Kumar Jain

Business Partner and Mentor.

Preface

Power tools have revolutionised the way we approach projects, from home improvement to professional construction. This book, **The Ultimate Guide to Power Tools: Master Efficiency, Ensure Safety, and Perfect Maintenance**, is born out of our combined 36 years of experience in the power tools industry. We have witnessed first-hand the evolution of these tools and the impact they have on efficiency, safety, and craftsmanship.

Our journey began with a legacy in hand tools, established by our late father, **Dr G.P. Garg,** and our late uncle, **Mr N.P. Garg,** in Old Delhi in 1980. Their dedication and entrepreneurial spirit laid the foundation for our family's success. In 2006, our father expanded into the power tools business and entrusted it to us, allowing us to build upon their legacy.

This book is a comprehensive guide designed to help both professionals and DIY enthusiasts master the use of power tools. It covers everything from selecting the right tools and mastering techniques to ensuring safety and maintaining equipment. Our goal is to provide you with the knowledge and confidence to tackle any project with precision and efficiency.

We hope this book serves as a valuable resource, empowering you to unlock the full potential of your power tools and achieve outstanding results in your projects.

Contents

Chapter 1:

Introduction to Power Tools

Overview of Power Tools

Power tools are mechanised devices designed to perform tasks with greater efficiency and less physical effort compared to hand tools. Unlike hand tools, which rely solely on human strength, power tools are driven by external power sources such as electricity, compressed air (pneumatic), or hydraulics. These tools are categorised based on their function, including cutting tools (e.g. circular saws, jigsaws), drilling tools (e.g. drills, impact drivers), grinding tools (e.g. angle grinders, bench grinders), and polishing tools (e.g. car polishers, sanders).

The Evolution of Power Tools

The history of power tools starts in the late 19th century with inventions like the electric drill by Arnot and Brain in 1889. Initially bulky and reliant on large motors, advances in motor tech made them smaller and more practical. Cordless tools became possible with Lithium-Ion batteries, offering mobility and ease.

Technological strides have enhanced power tools' efficiency, power, and user-friendliness. Brushless motors extend tool lifespan, while Lithium-Ion batteries enable longer run times and faster charging. Better materials like carbide tips and high-strength alloys improve performance, catering to both pros and DIYers with precision and reliability.

Safety standards have evolved, with regulations ensuring tools are safer to use. Features like automatic shutoffs, safety guards, and anti-kickback systems prevent accidents, boosting user confidence. Ergonomic design reduces fatigue and increases productivity, with vibration reduction tech, adjustable handles, and lightweight materials enhancing comfort and usability.

Consumer demand shapes power tool evolution, with trends driving new features and tech. DIY projects spur demand for versatile, user-friendly tools, while cordless options offer mobility and convenience. Multi-functional tools satisfy the need for efficiency and value, emphasising the industry's focus on meeting consumer preferences for innovation.

Tools Classification Based on Power Source

Corded Power Tools

Corded power tools are connected directly to an electrical outlet or power source via a power cord. They draw their power from the mains electricity supply, providing a consistent and reliable source of power. Corded tools are well-suited for heavy-duty, continuous-use applications as they are not limited by battery life. Examples include mains-powered drills, saws, grinders, and sanders.

Cordless Power Tools

Cordless power tools are powered by rechargeable batteries, eliminating the need for a power cord. This provides greater mobility and flexibility, allowing users to work in remote locations without access to an electrical outlet. Cordless tools are powered by Lithium-Ion (Li-ion) or nickel-cadmium (NiCad) batteries, which can be recharged when depleted. Examples include cordless drills, impact drivers, and circular saws.

Gasoline Power Tools

Gasoline power tools operate using internal combustion engines fueled by gasoline, making them entirely independent of power sources. These tools are particularly advantageous for tasks in remote locations where access to electricity or compressed air is limited. Gasoline-powered tools are well-known for their high power output and are typically used in heavy-duty applications such as forestry, landscaping, and construction in off-grid locations. Examples include chainsaws, brush cutter, and concrete saws.

Pneumatic Tools

Pneumatic tools are driven by compressed air rather than electricity. They use an air compressor to generate the high-pressure air that powers the tool's internal mechanisms. Pneumatic tools are well-suited for applications requiring high torque or force, such as impact wrenches, nail guns, and chipping hammers. They are often used in industrial and construction settings where compressed air is readily available.

Hydraulic Tools

Hydraulic tools use pressurised hydraulic fluid to transmit power and operate the tool. The hydraulic fluid is pumped through hoses to the tool, providing the necessary force and motion. Hydraulic tools are commonly used for heavy-duty applications that require significant power, such as jackhammers, rock drills, and hydraulic shears. They are often found in mining, construction, and demolition industries.

Hand Tools

Hand tools rely solely on human strength and skill for operation. These timeless implements have been essential to craftsmen and DIY enthusiasts for centuries, offering precision, control, and versatility without the need for external power sources. While they may require more physical effort than their powered counterparts, hand tools offer advantages in portability, durability and the ability to work in areas without access to electricity or compressed air.

Each power source has its own advantages and disadvantages in terms of power, portability, runtime, and suitability for different applications. Professionals and DIY users must carefully consider the specific requirements of their projects when selecting the most appropriate power tool and power source.

Why Power Tools Matter

Power tools have become an indispensable part of various industries, revolutionising work processes and delivering significant benefits. These versatile tools have transformed the way tasks are approached, enabling increased efficiency, precision and productivity across a wide range of applications.

Enhancing Efficiency and Productivity

One of the primary advantages of power tools is their ability to enhance efficiency and productivity. Compared to traditional hand tools, power tools are designed to expedite tasks, significantly reducing the time and effort required to complete them. Whether it's drilling, cutting, sanding, or shaping materials, power tools streamline processes, enabling professionals and DIY enthusiasts alike to accomplish more in less time.

Precision and Accuracy.

Power tools are engineered with precision in mind, ensuring accuracy in every task performed. The utilisation of advanced mechanisms and cutting-edge technology allows for meticulous cuts, perfectly aligned holes, and refined finishes. This precision not only saves time but also elevates the quality of the work produced, meeting professional standards and exceeding expectations.

Versatility and Adaptability.

Another compelling aspect of power tools is their versatility. From circular saws and drills to sanders and routers, these tools come in various types and sizes, catering to a wide array of tasks. Their adaptability allows users to tackle diverse projects with ease, making them indispensable in industries like construction, woodworking, metalworking, and even in everyday household repairs.

Empowering Professionals and DIY Enthusiasts.

The accessibility of power tools has empowered both professionals and DIY enthusiasts to explore their creativity and undertake complex projects with confidence. These tools enable individuals to tackle tasks that were once beyond their reach, from home improvement projects to intricate woodworking and crafting. The ease of use and versatility of power tools have made DIY projects more approachable and rewarding for beginners.

Revolutionising Work Processes

Power tools have fundamentally transformed work processes across various industries. They have streamlined construction operations, enhanced manufacturing efficiency, and enabled healthcare professionals to deliver more precise and effective treatments. By automating repetitive tasks, providing specialised features, and reducing physical strain, power tools have revolutionised the way work is done, leading to increased productivity, cost savings, and on-time project completion.

In conclusion, the importance of power tools cannot be overstated. These versatile tools have become the backbone of modern craftsmanship, empowering professionals and DIY enthusiasts alike to achieve greater efficiency, precision, and quality in their work. As technology continues to evolve, the role of power tools in revolutionising work processes will only become more pronounced, shaping the future of various industries and transforming the way we approach tasks.

Applications of Power Tools Across Various Fields

Construction and Renovation

Power tools are indispensable in construction and renovation, enabling efficient and precise execution of tasks across various projects. From residential buildings to large-scale infrastructure developments, power tools such as drills, saws, and demolition hammers are essential for tasks like drilling, cutting, and demolition. They facilitate faster construction timelines and higher quality finishes, contributing to the overall efficiency and success of construction projects.

Woodworking and Carpentry

In woodworking and carpentry, power tools have revolutionised traditional craftsmanship, allowing for greater precision and productivity. Essential tools like routers, planers, and sanders streamline processes such as shaping, smoothing, and finishing wood surfaces. With the aid of power tools, woodworkers can create intricate designs and achieve consistent results, enhancing the quality of their craftsmanship.

Metalworking

Power tools play a vital role in metalworking processes, facilitating tasks such as cutting, shaping, and welding with efficiency and accuracy. Tools like angle grinders, bench grinders, and rotary hammers are indispensable in metal fabrication and repair work, enabling the manipulation of metal materials with precision. Whether in industrial settings or artisan workshops, power tools are essential for shaping raw metal into finished products.

Automotive and Manufacturing

Power tools are integral to automotive repair and manufacturing processes, where precision and efficiency are paramount. Tools like impact wrenches, polishers, and air compressors are used extensively for tasks ranging from assembly and fabrication to maintenance and repair. Automation, combined with power tools, has streamlined manufacturing processes, allowing for increased productivity and consistency in output across the automotive and manufacturing sectors.

Home Improvement and DIY

The rise of DIY culture and home improvement projects has been greatly facilitated by the accessibility of power tools. DIY enthusiasts rely on tools like cordless screwdrivers, jigsaws, and pressure washers to tackle a wide range of projects, from simple repairs to complex renovations. Power tools empower individuals to take control of their living spaces, fostering creativity and self-sufficiency in home improvement endeavours.

Specialised Fields

Beyond traditional applications, power tools find utility in specialised fields such as landscaping, healthcare, and art. In landscaping, tools like brush cutters and chainsaws are used for land clearing and maintenance. In healthcare, surgical drills and other specialised tools aid medical professionals in procedures requiring precision and control. In art, rotary tools enable artists to sculpt and carve various materials, pushing the boundaries of creativity and expression in artistic endeavours.

Chapter 2:

Essential Power Tools: A Detailed Guide

Power tools come in various types and serve a multitude of functions across different fields. This chapter provides a detailed overview of the most commonly used power tools, organised into specific categories based on their primary function. Each category includes a list of tools, along with descriptions of their features and typical applications. This structured approach helps you easily find and understand the tools you need for any task.

Understanding the various power tools, their functions, and uses can significantly enhance efficiency and precision in a wide range of tasks, from construction and woodworking to metalworking and home improvement. This guide aims to provide a comprehensive overview of essential power tools, helping users select the right tool for their needs and applications.

This guide offers detailed descriptions of various power tools, their components and their functions. It is designed to help readers, especially those new to using power tools, understand each tool's purpose and operation.

CUTTING TOOLS

Tools Designed for Cutting Various Materials, Including Metal, Wood, and Masonry.

Angle Grinder

A handheld power tool equipped with a rotating abrasive disc, used for cutting, grinding, and polishing various materials.

Uses:

- Cutting metal, masonry, and tile; grinding welds; smoothing rough surfaces; and removing rust or paint.

Attachments:

- Grinding discs for grinding and smoothing surfaces.
- Cutting wheels for cutting through metal, stone, and concrete.
- Wire brushes for removing rust, paint, and surface contaminants.

Technical Details:

Disc sizes typically range from 4" to 9".

Speed ratings from 6,000 to 11,000 RPM.

Compatible with various materials, including metal, stone, and concrete.

Step-by-Step Guide:

1. Select the appropriate disc for the task.
2. Secure the workpiece firmly.
3. Hold the grinder with both hands and maintain a firm grip.
4. Start the grinder before contacting the workpiece.
5. Apply steady, moderate pressure while moving the grinder smoothly.
6. Allow the tool to cool down between cuts.

Expert Advice:

Common Mistake: Applying too much pressure, leading to disc damage or kickback.

Solution: Let the tool do the work; apply gentle, consistent pressure.

Usage Tip: For smoother cuts, use sharp blades.

Circular Saw

A power tool featuring a circular blade that rotates to cut materials, primarily wood, metal and plastic.

Uses:

- Making straight cuts (crosscuts and rip cuts) in lumber and sheet materials, like plywood and particleboard.

Blade Types:

- Carbide-tipped blades for wood and composite materials.
- High-speed steel (HSS) blades for metal.
- Abrasive blades for cutting concrete, stone, and other masonry materials.

Technical Details:

Blade sizes typically range from 5″ to 7″.

Cutting depths vary from 1.5″ to 2.5″ at 90 degrees.

Step-by-Step Guide:

1. Measure and mark your cutting line.
2. Adjust blade depth to slightly exceed the material thickness.
3. Secure workpiece to prevent movement.
4. Align the saw blade with the cutting line.
5. Start saw before contacting material.
6. Guide saw steadily along the cutting line.

Expert Advice:

Common Mistake: Forcing the saw through the material.

Solution: Let the saw's weight and sharp blade do the cutting.

Usage Tip: For cleaner cuts, use a zero-clearance insert or a guide.

Core Drill

A specialised drilling tool used to create large-diameter holes in materials such as concrete, stone, and walls.

Uses:

Creating openings for pipes, cables, and conduits in construction, plumbing, electrical, and AC installations.

Technical Details:

Core bit diameters range from 50mm to 120mm, or larger.

Typically operate at lower speeds (up to 2100 RPM) compared to standard drills.

Step-by-Step Guide:

1. Mark the centre of the hole to be drilled.
2. Secure the core drill stand firmly to the work surface.

3. Attach the appropriate core bit for the material and hole size.
4. Set up water supply for cooling, if applicable.
5. Start drilling at low speed, gradually increasing as the bit bites.
6. Maintain steady pressure and allow the drill to do the work.

Expert Advice:

Common Mistake: Applying too much pressure can lead to bit damage or motor strain.

Solution: Use steady, moderate pressure and let the bit's weight assist in cutting.

Usage Tip: For cleaner exit holes, slow down as you near breakthrough.

Cut Off Machine

A power tool designed for cutting hard materials, like metal.

Uses:

- Cutting metal pipes, rods, bars, and rebar in construction, metalworking, and fabrication industries.

Blade Types:

- Abrasive wheels for ferrous and non-ferrous metals.
- Diamond blades for special requirements.

Technical Details:

Blade sizes typically range from 12" to 16".

Cutting capacities vary from 4" to 6" depending on the model.

Speed ratings from 3,000 to 4,000 RPM.

Step-by-Step Guide:

1. Secure material firmly in the machine's vise
2. Select the appropriate blade for the material.
3. Lower guard and ensure proper alignment.
4. Start the machine and allow the blade to reach full speed.
5. Lower the blade slowly into the material, maintaining steady pressure.
6. Complete the cut and allow the blade to stop before raising

Expert Advice:

Common Mistake: Cutting materials not suited for the machine's capacity.

Solution: Always check material specifications against machine capabilities.

Usage Tip: For longer blade life, avoid overpressure.

Marble Cutter

A specialised power tool for cutting stone and tiles, particularly marble, granite, and ceramic.

Uses:

- Cutting tiles to fit around corners, edges, and irregular shapes in tile laying and stone masonry.

Blade Types:

Diamond blades for precision and durability in cutting hard materials like marble, granite, and ceramic.

Technical Details:

Blade sizes are typically 4″ to 6″.

Cutting depths range from 1″ to 2″.

Water-feed systems for cooling and dust suppression.

Step-by-Step Guide:

1. Mark cutting line on marble or tile
2. Fill the water reservoir or connect to the water supply.
3. Align the blade with the cutting line.
4. Start saw and allow water to flow.
5. Guide saw steadily along the line, maintaining a consistent speed.
6. Complete the cut and allow the blade to stop before setting it down

Expert Advice:

Common Mistake: Cutting without proper water flow can lead to blade damage and dust.

Solution: Ensure consistent water flow throughout the cutting process.

Usage Tip: For complicated cuts, use a wet tile saw for greater precision.

Miter Saw

A power tool with a circular blade that pivots on a vertical axis to make precise crosscuts and miter cuts.

Uses:

- Cutting angles, miters, bevels and compound cuts in woodworking and carpentry.

Technical Details:

Blade sizes are typically 8" to 12".

Miter ranges are usually 0-45 degrees left and right.

Bevel capacities are often up to 45 degrees.

Step-by-Step Guide:

1. Set desired miter and bevel angles
2. Secure workpiece against the fence.
3. Lower blade guard and align it with the cutting mark.
4. Start the saw and allow it to reach full speed.
5. Lower the blade smoothly through the material.
6. Release the trigger and allow the blade to stop before raising

Expert Advice:

Common Mistake: Cutting warped or twisted boards can lead to inaccurate cuts.

Solution: Use support stands to properly support long boards.

Usage Tip: For repetitive cuts, use stop blocks to ensure size consistency.

Slab Cutter

A power tool for cutting concrete and large stone slabs into smaller pieces.

Uses:

- Cutting slabs of materials like marble, granite, and concrete for countertops, floors, and paving.

Technical Details:

Blade sizes typically 14" to 16"

Cutting depths up to 6", depending on the model.

Often features water-cooling systems for dust suppression.

Step-by-Step Guide:

1. Measure and mark cutting line on slab
2. Position the slab on the cutting table or secure work area.
3. Align the blade with the cutting mark

4. Start saw and engage the water-cooling system
5. Push the saw steadily through the material.
6. Complete the cut and allow the blade to stop before moving the slab.

Expert Advice:

Common Mistake: Cutting without proper support can lead to material breakage.

Solution: Ensure full support of the slab throughout the cutting process.

Usage Tip: For cleaner edges, use a slower feed rate on the final pass.

Wall Chaser

A power tool designed to cut channels or grooves into walls and other surfaces.

Uses:

Installing electrical wiring, plumbing pipes and conduits within walls in construction and renovation projects.

Technical Details:

Blade diameters are typically 4″ to 6″.

Cutting depths up to 1.5″.

Adjustable cutting widths to accommodate various channel sizes.

Step-by-Step Guide:

1. Mark channel location on wall
2. Adjust blade depth and width as needed.
3. Connect dust suppression water system.

4. Start the tool and plunge carefully into the wall.
5. Guide the chaser along the marked line with steady pressure.
6. Make multiple passes if a deeper channel is required.

Expert Advice:

Common Mistake: Cutting without proper dust control, leading to excessive dust.

Solution: Always use a proper dust suppression system.

Usage Tip: For cleaner cuts, score the outline of the channel first, then remove material.

DRILLING & DRIVING TOOLS

Tools Used for Creating Holes and Driving Screws into Various Materials.

Types:

- Corded drills are for continuous use and powered by electricity.
- Cordless drills for portability and convenience.
- Hammer drills with a hammering action for masonry and concrete.

Drill Machine

A versatile power tool designed to create holes by rotating a cutting tool (drill bit).

Uses:

- Drilling holes in concrete, wood, and metal in construction, woodworking, and metalworking.

Technical Details:

Chuck sizes typically range from 6.5mm to 13mm.

Speed ranges from 0 to 4000 RPM for most models.

Some models offer left-right rotation control.

Step-by-Step Guide:

1. Select appropriate drill bit for the material and hole size
2. Secure the drill bit in the chuck and tighten.
3. Mark the drilling location and set the drill speed.

4. Hold the drill perpendicular to the surface.
5. Apply steady pressure and start drilling.
6. Back out periodically to clear debris.

Expert Advice:

Common Mistake: Using excessive pressure can lead to bit breakage or motor strain.

Solution: Let the drill do the work; use steady, moderate pressure.

Usage Tip: For precise hole placement, use a centre punch to create a starting point.

Impact Drill

A drill designed for tough materials, like concrete and stone, combining rotary with repetitive impact action.

Uses:

- Heavy-duty applications, where standard drills struggle to penetrate hard surfaces.

Features:

- Variable speed control and impact settings for different materials.

Technical Details:

Impact rates typically range from 0 to 48,000 BPM (Blows Per Minute).

Chuck sizes are usually 10 mm or 16 mm.

Impact mode can be set to On or Off.

Step-by-Step Guide:

1. Select appropriate drill bit (preferably impact-rated)
2. Set drill to hammer mode for masonry or concrete.
3. Mark drilling location and set speed.
4. Apply firm, steady pressure while drilling.
5. Use short bursts for harder materials.
6. Clean out hole periodically to prevent bit binding.

Expert Advice:

Common Mistake: Using low-quality bits in impact mode, leading to premature wear.

Solution: Always use impact-rated quality bits for maximum durability.

Usage Tip: For larger holes in concrete, start with a smaller pilot hole.

Cordless Screwdriver

A handheld tool for driving screws quickly and efficiently.

Uses:

- Furniture assembly, electrical work, DIY projects, and general repairs.

Advantages:

Portability and ease of use reduce fatigue during extended use.

Technical Details:

- Voltage ranges around 12V for most models.
- Chuck sizes are typically 10 mm.
- Adjustable Torque

Step-by-Step Guide:

1. Select appropriate bit for screw head
2. Insert bit into chuck.
3. Adjust torque setting.
4. Align the bit with the screw head
5. Apply moderate pressure and engage the trigger.
6. Release the trigger when the screw is fully seated or released.

Expert Advice:

Common Mistake: Over-tightening screws, leading to stripped heads or damaged material.

Solution: Use appropriate torque settings and stop when the screw is flush.

Usage Tip: For delicate materials, start screws by hand before using the machine.

Electric Screwdriver

A power tool for driving screws quickly and efficiently, often used in manufacturing, assembly lines and DIY projects.

Uses:

- High-volume screwdriving tasks, where speed and efficiency are essential.

Advantages:

- Corded electric screwdrivers for continuous power,

Technical Details:

- Chuck sizes are typically 10 mm.
- Adjustable Torque,
- Max speed around 1000 rpm.

Step-by-Step Guide:

1. Select appropriate bit for screw type
2. Set the desired speed and torque (if adjustable).
3. Position screw on the bit tip.
4. Align the screw with the desired location.
5. Apply steady pressure and engage the trigger.
6. Release the trigger when the screw is fully seated.

Expert Advice:

Common Mistake: Using the incorrect bit size, leading to stripped screws.

Solution: Always match the bit size and shape precisely to the screw head.

Usage Tip: For repetitive tasks, use a screw guide or magnetic bit holder for faster work.

Rotary Hammer

A heavy-duty power tool designed for drilling and chiselling in concrete, masonry, and stone.

Uses:

- Drilling anchor holes, breaking up concrete, and chiselling away materials in construction and demolition.

Features:

- Multiple modes (drilling, hammer drilling, chiselling).
- SDS chuck systems for easy bit changes.

Technical Details:

SDS chuck size range, 20mm to 26mm.

Speeds range from 0 to 1500 RPM, with 0 to 5000 BPM.

Step-by-Step Guide:

1. Select appropriate bit for task (drill, chisel, or hammer)
2. Insert bit into SDS chuck.

3. Set mode (drill, hammer drill, or chisel).
4. Mark drilling/chiselling location.
5. Apply firm, steady pressure while operating.
6. Use a dust extraction system for cleaner operation.

Expert Advice:

Common Mistake: Using excessive force can lead to premature tool wear.

Solution: Let the tool's weight and impact action do the work.

Usage Tip: For drilling larger holes, start with a smaller pilot hole.

Bench Drill

A stationary power tool mounted on a bench or stand for precision drilling.

Uses:

- Accurate and repetitive drilling tasks in metalworking, woodworking, and manufacturing.

Features:

Adjustable tables, variable speed, and depth stop for precise control.

Technical Details:

Chuck sizes typically range from 10mm to 13mm.

Speeds are usually adjustable from 500 to 3000 RPM.

Table sizes are around 8" x 12".

Step-by-Step Guide:

1. Secure workpiece to drill table
2. Insert the appropriate drill bit into the chuck.
3. Adjust table height and drill depth stop.
4. Set an appropriate speed for the material and bit size.
5. Lower quill to begin drilling.
6. Raise quill periodically to clear chips.

Expert Advice:

Common Mistake: Failing to secure the workpiece, leading to injury or inaccurate holes.

Solution: Always use clamps or a drill vise to secure workpieces.

Usage Tip: For precise depth control, use the depth stop feature.

Drill Stand

A stationary tool accessory designed to hold a drill machine securely for stability and precision.

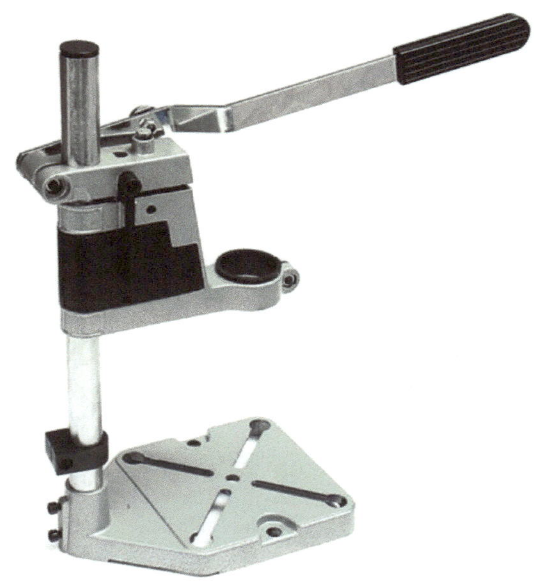

Uses:

- Precision drilling tasks, where accuracy and repeatability are crucial, such as in metalworking, woodworking, and model-making.

Features:

Adjustable height and angle settings, and a stable base, ensure consistent results.

Technical Details:

Compatible with most handheld drills.

Typically, accommodate drill base diameters up to 13mm.

Vertical travel ranges are normally 3".

Step-by-Step Guide:

1. Secure drill stand to work surface
2. Mount drill in stand's collar.
3. Secure workpiece beneath the drill.
4. Lower drill using a lever mechanism.
5. Raise the drill after completing the hole.

Expert Advice:

Common Mistake: Inadequate stabilisation of the drill stand.

Solution: Ensure the stand is firmly bolted or clamped to a stable surface.

Usage Tip: Use a drill press vise with the stand for even greater precision.

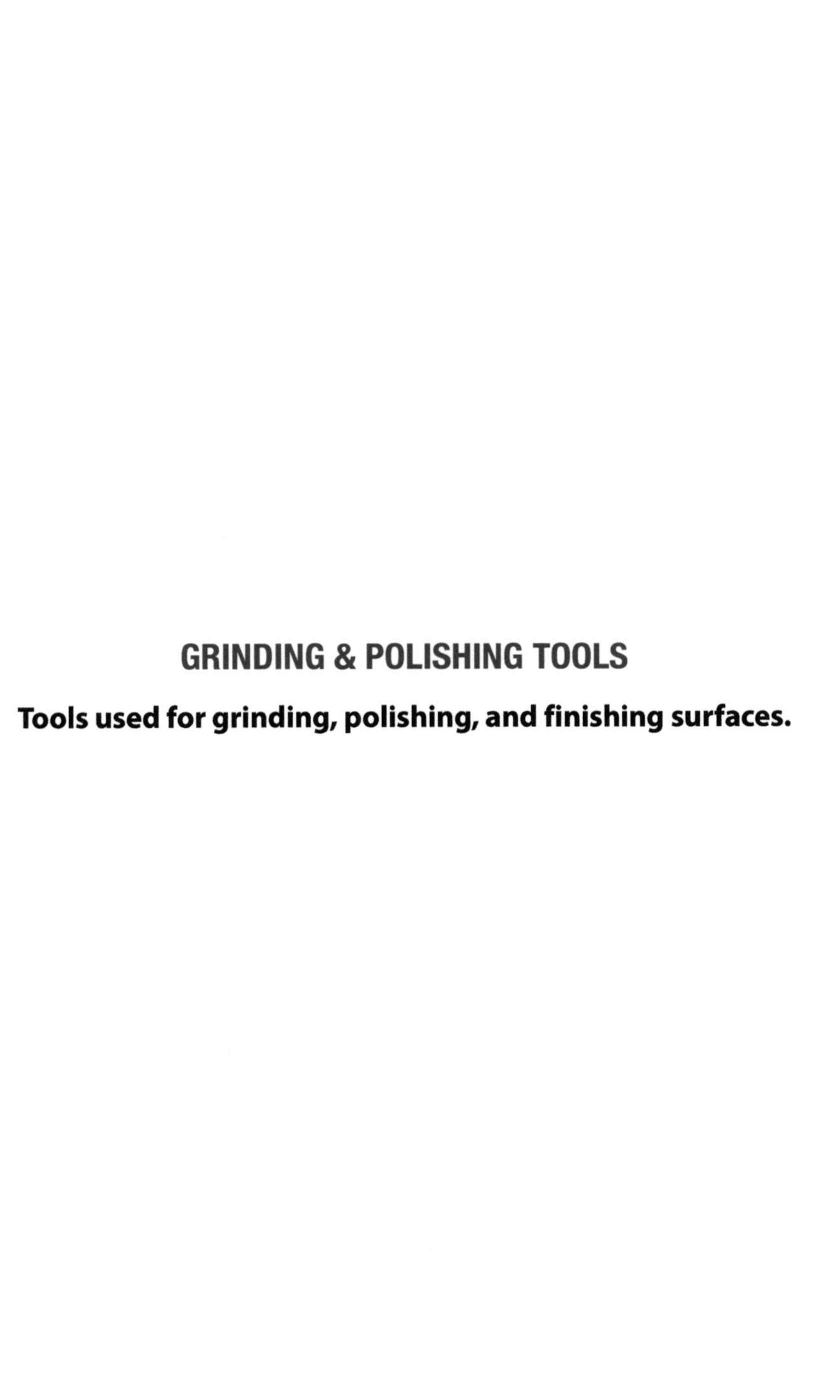

GRINDING & POLISHING TOOLS

Tools used for grinding, polishing, and finishing surfaces.

Angle Grinder

A versatile power tool used for cutting, grinding and polishing materials.

Uses:

Metal cutting, masonry cutting, surface preparation, and rust removal.

Attachments:

- Grinding discs for metal surfaces.
- Cutting wheels for metal and masonry.
- Wire brushes for rust and paint removal.

Technical Details:

Disc sizes typically range from 4" to 9".

Speed ratings from 6,000 to 11,000 RPM.

Step-by-Step Guide:

1. Select appropriate disc for the task (cutting, grinding, or polishing)

2. Secure the workpiece firmly.
3. Hold the grinder with both hands, maintaining a firm grip.
4. Start the grinder before contacting the workpiece.
5. Apply steady, moderate pressure while moving the grinder smoothly.
6. Allow the tool to cool down between uses.

Expert Advice:

Common Mistake: Using a worn-out or damaged disc.

Solution: Inspect discs before each use and replace if worn or damaged.

Usage Tip: For safer operation, use the grinding wheel guard.

Angle Grinder Stand

A tool accessory that converts a handheld angle grinder into a benchtop tool for stationary tasks.

Uses:

- Precision cutting and grinding, where stability and accuracy are essential.

Features:

Adjustable settings for angle and height.

Technical Details:

Compatible with most angle grinders.

Some models feature built-in spark guards and work rests.

Step-by-Step Guide:

1. Securely mount the angle grinder in the stand
2. Ensure all locking mechanisms are engaged.

3. Place the workpiece on the work rest.
4. Start the grinder and carefully bring the workpiece to the disc.
5. Apply steady pressure and move the workpiece as needed.

Expert Advice:

Common Mistake: Neglecting to secure the grinder properly in the stand.

Solution: Double-check all mounting points and test stability before use.

Usage Tip: Select the appropriate disk for the selected task.

Bench Grinder

A stationary power tool mounted on a bench or pedestal for sharpening tools and grinding metal.

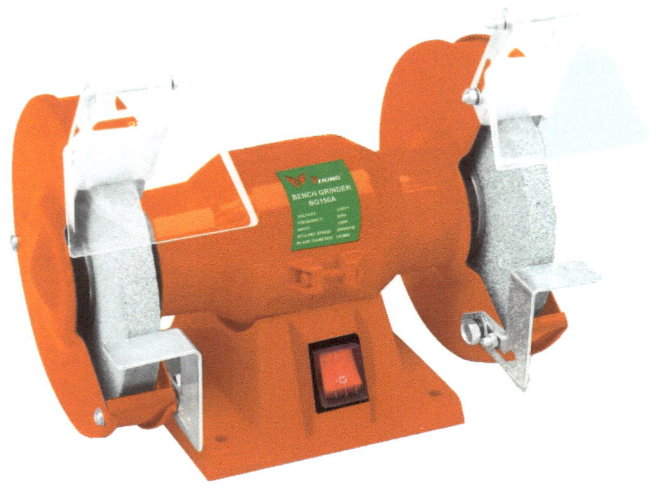

Uses:

- Sharpening chisels, drill bits, mower blades and grinding metal surfaces.

Attachments:

- Grinding wheels for general grinding and sharpening.
- Wire brushes for removing rust and paint.
- Buffing wheels for polishing metal surfaces.

Technical Details:

Wheel sizes are typically 6″ to 8″.

Speed ratings are usually between 3,000 and 3,600 RPM.

Step-by-Step Guide:

1. Adjust tool rests within reach of the wheel
2. Ensure eye shields are properly positioned.
3. Turn on the grinder and allow it to reach full speed.
4. Hold the workpiece firmly against the tool rest.
5. Move the workpiece across the face of the wheel.
6. Dip the workpiece in water frequently to prevent overheating.

Expert Advice:

Common Mistake: Using too much pressure, causing wheel glazing or workpiece burning.

Solution: Use light, consistent pressure and keep the workpiece moving.

Usage Tip: Swarf (clean) the wheels regularly to maintain optimal grinding performance.

Car Polisher

A power tool used for polishing and buffing automotive surfaces to restore shine and remove imperfections.

Uses:

- Car detailing, paint correction, and applying wax or sealant.

Attachments:

- Polishing pads for applying polish and wax.
- Buffing pads (Softer) for removing polish residue and achieving a high-gloss finish.

Technical Details:

Pad sizes are typically 7".

Speed ranges from 600 to 3,000 RPM for variable speed models.

Step-by-Step Guide:

1. Wash and dry the vehicle thoroughly
2. Apply polishing compound to the pad or surface.

3. Start the polisher before contacting the surface.
4. Move the polisher in overlapping patterns, applying light pressure.
5. Work in small sections, frequently checking results.
6. Clean the pad or switch to a clean pad as needed.

Expert Advice:

Common Mistake: Using too high-speed or too much pressure, causing swirl marks.

Solution: Start with lower speeds and light pressure, increasing as needed.

Usage Tip: For best results, follow polishing with a separate waxing step.

WOOD WORKING TOOLS

Tools Specifically Designed for Working with Wood, Including Cutting, Shaping and Finishing.

Router

A tool used for routing or hollowing out an area in the face of a piece of wood, plastic, or other materials.

Uses:

- Shaping edges, creating joinery, and hollowing out wood.

Technical Details:

- Speed ranges from 8,000 to 25,000 RPM.

Step-by-Step Guide:

1. Select and install the appropriate router bit.
2. Adjust the cutting depth.
3. Secure the workpiece.
4. Turn on the router and let it reach full speed.
5. Guide the router along the desired path.

Expert Advice:

- Common Mistake: Moving the router too quickly.
- Solution: Move at a steady, controlled pace.
- Usage Tip: Use a router table for more precise control.

Planer

A tool used to smooth and flatten wood surfaces, reducing the thickness of the material.

Uses:

- Smoothing rough lumber, resizing boards, and creating uniform thickness.

Technical Details:

- Cutting width ranges from 82mm to 90mm.
- Cutting depth per pass is typically up to 3 mm.

Step-by-Step Guide:

1. Adjust the planer to the desired thickness.
2. Secure the workpiece.
3. Feed the wood through the planer.
4. Make multiple passes until the desired thickness is achieved.

Expert Advice:

- Common Mistake: Removing too much material in one pass.
- Solution: Make multiple shallow passes.
- Usage Tip: Check for nails or screws before planning.

Jig Saw

A power tool with a reciprocating blade used for cutting intricate shapes and curves in wood, metal, and other materials.

Uses:

- Cutting curves, angles and shapes in various materials.

Technical Details:

- Speed ranges from 500 to 3,500 strokes per minute (SPM).
- Blade types include T-shank (newer models) and U-shank (vintage models).

Step-by-Step Guide:

1. Select and install the appropriate blade.
2. Secure the workpiece.
3. Mark the cutting line.

4. Start the saw and guide it along the line.
5. Adjust the speed as needed for different materials.

Expert Advice:

- Common Mistake: Forcing the saw through the material.
- Solution: Let the blade do the work.
- Usage Tip: Use a fine-toothed blade for smoother cuts.

Rotary Sander:

A power tool with a circular sanding disc that rotates at high speeds for aggressive material removal and shaping.

Uses:

- Heavy-duty sanding, stripping paint or varnish, levelling uneven surfaces, and shaping wood or metal in construction, automotive work, and metalworking.

Technical Details:

- Disc sizes are typically 5".
- Rotation speeds generally between 2,000 and 3,500 RPM (Revolutions Per Minute).

Step-by-Step Guide:

1. Select appropriate grit and type of sanding disc for the task.
2. Securely attach the sanding disc to the tool.
3. Set up proper dust collection, or wear appropriate respiratory protection.

4. Start the sander before making contact with the surface.
5. Apply light pressure and move the sander in a consistent direction.
6. Regularly check the surface and replace the disc as needed.

Expert Advice:

Common Mistake: Applying too much pressure, which can cause deep gouges or burn marks.

Solution: Let the tool's weight and rotation do the work, applying only light pressure.

Usage Tip: For smoother results on wood, sand with the grain direction, when possible.

Orbital Sander:

A power tool with a sanding pad that moves in small circular orbits for smooth sanding.

Uses:

- Finishing and smoothing surfaces, removing paint or varnish, and preparing surfaces for painting or staining in woodworking and home improvement.

Technical Details:

Pad sizes are typically 90 mm x 190 mm.

Orbital speeds range from 8,000 to 12,000 OPM (Orbits Per minute).

Step-by-Step Guide:

1. Choose appropriate grit sandpaper for the task
2. Attach sandpaper to the sanding pad.
3. Ensure a proper dust collection setup.
4. Start the sander before contacting the surface.

5. Move the sander in overlapping circular or figure-eight patterns.
6. Periodically check the surface and replace sandpaper as needed.

Expert Advice:

Common Mistake: Tilting the sander, leading to uneven sanding or swirl marks.

Solution: Keep the sander flat against the surface at all times.

Usage Tip: For fine finishing, use a random orbital sander for a swirl-free finish.

CLEANING TOOLS

Tools That Assist in Cleaning and Maintaining Surfaces and Environments.

Pressure Washer:

A pressure washer is a high-pressure mechanical sprayer used for cleaning surfaces by forcefully removing dirt, grime, and other contaminants. It consists of a motor (electric or gasoline), a pump, and a high-pressure hose.

Uses:

Cleaning driveways, sidewalks, decks, vehicles, outdoor furniture, and building exteriors.

- Removing mould, mildew and tough stains from various surfaces.

Preparing surfaces for painting or staining by removing old paint, dirt, and debris.

Features:

Adjustable jet settings for different cleaning tasks.

- Various nozzles and attachments for specific cleaning applications,

- Detergent injection systems for enhanced cleaning performance.

Technical Details:

Pressure ratings typically range from 80 to 120 bar.

Water flow rates from 7 to 10 LPM (Litres Per Minute)

Power sources include electricity and battery-powered.

Step-by-Step Guide:

1. Connect water supply and ensure proper flow
2. Attach the appropriate nozzle for the task.
3. Start the pressure washer and purge air from the system.
4. Begin spraying at a distance, moving closer as needed.
5. Use sweeping motions to clean surfaces evenly.
6. Apply detergent (if needed) and rinse thoroughly.

Expert Advice:

Common Mistake: Using too high-pressure or holding nozzle too close to surfaces.

Solution: Start with lower pressure and a wider nozzle, adjusting as needed.

Usage Tip: For stubborn stains, use a rotating or turbo nozzle for increased cleaning power.

Wet and Dry Vacuum Cleaner

A wet and dry vacuum cleaner is a versatile cleaning tool capable of removing both liquid spills and dry debris. It consists of a motor, a vacuum canister, and various attachments for different cleaning tasks.

Uses:

- Cleaning up wet spills, dry dirt, sawdust and other debris in construction sites, workshops and household settings.
- Handling post-renovation cleanup by removing fine dust and construction debris.
- Vacuuming liquid spills, unclogging drains and cleaning up after floods or leaks.

Features:

- Versatility to handle different types of cleaning tasks.
- Powerful suction capabilities for efficient cleanup.
- Various attachments for specific cleaning requirements, such as crevice tools, brushes and extension wands.
- Durable construction, with a large-capacity canister, for heavy-duty use.

Easy-to-clean filters and tanks for maintenance and longevity.

Technical Details:

Tank capacities typically range from 15 to 50 litres.

Suction power varies from size to size.

Step-by-Step Guide:

1. Choose appropriate filter for wet or dry use
2. Attach a suitable nozzle or brush for the task.
3. Turn on the vacuum and begin cleaning.
4. For wet pickup, empty the tank when full.
5. For dry pickup, check and clean the filter regularly.
6. After use, clean and dry all components thoroughly.

Expert Advice:

Common Mistake: Using paper filters for wet pickup.

Solution: Always use the appropriate filter for wet or dry applications.

Usage Tip: For improved suction on hard surfaces, use a squeegee attachment.

Air Blower

An air blower is a device designed to blow air at high speeds, typically consisting of a motor-driven fan or impeller that generates airflow and directs it through a nozzle or outlet.

Uses:

- Cleaning debris from surfaces, such as workshop floors, computer keyboards or outdoor spaces.
- Drying surfaces after cleaning or washing, speeding up processes such as car drying, or drying freshly painted surfaces.
- Inflating objects such as inflatable mattresses, sports balls, or inflatable pool toys.

Technical Details:

Air volume typically ranges from 2 to 3 m^3/min (Cubic Metres per minute).

Variable speed

Power sources include electric motors and battery-powered.

Step-by-Step Guide:

1. Attach appropriate nozzle
2. Clear the area of loose objects that could become projectiles.
3. Start the blower and adjust speed if a variable control is available.
4. Begin blowing from one end of the area, working systematically.
5. Use sweeping motions to direct debris.
6. For dusty areas, consider wearing eye and respiratory protection.

Expert Advice:

Common Mistake: Using a blower in enclosed spaces can spread dust.

Solution: In enclosed areas, use vacuums or dust collection systems instead.

Usage Tip: For more precise control, use speed control and concentrator nozzle attachment.

GARDEN TOOLS

Tools Used for Various Gardening and Landscaping Tasks.

Electric Chainsaw

A portable power tool with a motor-driven chain looped around a guide bar, primarily used for cutting wood.

Uses:

- Tree felling, limbing, bucking, and pruning in forestry, landscaping, and residential settings.

Advantages:

- Reduced noise and lower maintenance requirements compared to petrol chainsaws.

Technical Details:

Chain bar lengths typically range from 12" to 18".

Step-by-Step Guide:

1. Check chain tension and oil level
2. Clear the work area and plan the cut.
3. Grip the saw firmly with both hands.
4. Start saw and allow the chain to reach full speed.

5. Begin cutting using the lower part of the bar.
6. Maintain steady pressure, letting the saw do the work.

Expert Advice:

Common Mistake: Improper chain tension, leading to poor performance or safety issues.

Solution: Regularly check and adjust chain tension according to the machine.

Usage Tip: For precision cuts, use the bumper spike as a pivot point.

Petrol Chainsaw

A heavy-duty cutting tool powered by a two-stroke and four-stroke petrol engine, featuring a chain with sharp teeth for cutting through wood.

Uses:

- Tree felling, pruning branches, and bucking logs in forestry and logging operations.

Maintenance:

- Regular maintenance, including chain sharpening, tension adjustment, air filter cleaning, and engine servicing, is essential.

Technical Details:

Engine sizes typically range from 30cc to 60cc.

Bar lengths from 14" to 24" or longer.

Two-stroke engines require an oil-gas mixture, while four-stroke engines use separate oil reservoirs.

Step-by-Step Guide:

1. Check fuel, bar oil, and chain tension
2. Clear the work area and plan an escape route.

3. Start saw on the ground or with secure footing.
4. Allow the saw to warm up before cutting.
5. Use proper cutting techniques for felling, limbing, or bucking.
6. Maintain the saw between cuts, checking for heat buildup.

Expert Advice:

Common Mistake: Improper fuel mixture leading to engine damage.

Solution: Always use the correct fuel-oil ratio specified for the machine.

Usage Tip: For safer felling, use felling wedges to control the direction of the fall.

Earth Auger

A tool used for drilling holes in the ground for posts, trees and other installations.

Uses:

- Drilling holes for fence posts, planting trees, and installing signs.

Technical Details:

- Auger diameters typically range from 2 to 12 inches.
- Power sources include gas, electric, and manual.

Step-by-Step Guide:

1. Select the appropriate auger bit.
2. Attach the bit to the auger.

3. Start the tool and position it over the desired location.
4. Apply steady pressure to drill the hole.
5. Clear the hole of debris, as needed.

Expert Advice:

- Common Mistake: Forcing the auger through hard soil.
- Solution: Let the tool do the work; use a steady, controlled pace.
- Usage Tip: For deeper holes, use extension rods.

Brush Cutter

A tool used for cutting through dense vegetation, thick grass and small plants.

Uses:

Clearing overgrown areas, cutting thick grass, and trimming bushes.

Technical Details:

- Cutting widths range from 10 to 15 inches.
- Power sources include gas, electric, and battery-operated.

Step-by-Step Guide:

1. Select the appropriate cutting attachment.
2. Adjust the height and angle of the cutter.
3. Secure the work area.
4. Start the cutter and guide it through the vegetation.
5. Maintain a steady pace for even cutting.

Expert Advice:

- Common Mistake: Cutting too close to the ground.
- Solution: Keep the cutter head level.
- Usage Tip: Use a backpack (harness) for better control and reduced fatigue.

MISCELLANEOUS TOOLS

A diverse range of tools used for specialised tasks across various trades.

Air Compressor

An air compressor converts power (usually from an electric motor, diesel engine, or gasoline engine) into potential energy stored in pressurised air. It comprises a compressor unit, a tank to store pressurised air, and various controls and valves to regulate airflow.

Uses:

- Powering pneumatic tools, such as air drills, impact wrenches, nail guns, and paint sprayers, in industrial, automotive, and construction settings.
- Inflating tyres, inflatable toys, sports equipment and pneumatic mattresses.
- Providing a consistent source of compressed air for spray painting results in smooth and even finishes.

Technical Details:

Tank sizes typically range from 30 to 300 litres.

Pressure ratings from 90 to 175 PSI.

Horsepower ratings from 1 to 10 HP for most models.

Step-by-Step Guide:

1. Ensure proper ventilation in the work area
2. Check oil level (if applicable) and drain moisture from the tank.
3. Connect the air hose and the desired pneumatic tool.
4. Start the compressor and allow it to build pressure.
5. Adjust the regulator to the appropriate pressure for the tool.
6. Use the tool, allowing the compressor to cycle on and off as needed.

Expert Advice:

Common Mistake: Neglecting to drain the tank, leading to rust and decreased performance.

Solution: Drain the tank after each use, or at least weekly.

Usage Tip: Use a filter/regulator/lubricator (FRL) unit to protect tools and improve air quality.

Paint Sprayer

A tool that atomises paint into a fine mist for even application on surfaces.

Uses:

Painting large surfaces, applying even coats of paint, and achieving smooth finishes in home improvement, automotive painting, and industrial applications.

Technical Details:

Spray patterns typically include round, horizontal and vertical.

Flow rates from 0.1 to 0.5 gallons per minute.

Tip sizes range from 0.009" to 0.025" for most applications.

Step-by-Step Guide:

1. Thin paint according to manufacturer's instructions
2. Set up the sprayer and test the spray pattern on cardboard.

3. Adjust pressure and tip size as needed.
4. Start spraying off the surface, then move across in even strokes.
5. Overlap each pass by 50% for even coverage.
6. Clean the sprayer thoroughly after use.

Expert Advice:

Common Mistake: Spraying too close or too far from the surface.

Solution: Maintain a consistent distance of 6-12 inches from the surface.

Usage Tip: For large areas, use an extension pole to reduce fatigue and maintain consistent coverage.

Heat Gun

A tool that emits a stream of hot air for various heating applications.

Uses:

- Stripping paint, shrinking heat shrink tubing, bending plastics, and thawing frozen pipes in home improvement and industrial settings.

Features:

Adjustable temperature settings and airflow control for different applications.

Technical Details:

Temperature ranges typically from 100°C to 600°C.

Airflow rates from 7 to 17 CFM (Cubic Feet per minute).

Power ratings from 1000 to 2000 watts.

Step-by-Step Guide:

1. Select appropriate temperature and airflow settings
2. Test on a small, inconspicuous area first.
3. Hold the heat gun 2-6 inches from the surface.
4. Move the gun constantly to avoid overheating one spot.
5. Use a scraper or other tools as needed for paint removal.
6. Allow the surface to cool before touching or applying new finishes.

Expert Advice:

Common Mistake: Using excessive heat can lead to surface damage or fire hazards.

Solution: Start with lower heat settings and increase gradually as needed.

Usage Tip: For paint stripping, use a wide nozzle attachment for more even heat distribution.

Arc Welding Machine

An arc welding machine is a power tool designed to fuse metal pieces together by creating a strong bond through fusion. It operates by generating an electric arc between an electrode and the metal, melting the materials to form a weld.

Uses

- Commonly employed in construction, manufacturing, automotive repair, and metal fabrication for tasks such as welding, cutting, and brazing.
- Ideal for creating robust joints in metal structures, repairing metal parts and fabricating custom metal components.

Advantages of Arc Welding Machines

- Cost-effective and versatile machines capable of welding various metals using consumable or non-consumable electrodes.

- Portable: Suitable for heavy-duty and outdoor welding tasks.
- Can handle dirty or rusty metals more effectively than MIG or TIG welders.

Technical Details:

Amperage ranges typically from 200 to 350 amps.

Duty cycles from 20% to 60% at maximum amperage.

For most applications, electrode sizes used are 2.4mm to 4mm.

Step-by-Step Guide:

1. Prepare metal surfaces by cleaning and bevelling edges if necessary
2. Set amperage according to electrode size and metal thickness.
3. Attach ground clamp to the workpiece or welding table.
4. Strike an arc by tapping or scratching the electrode on the metal.
5. Maintain a consistent arc length and travel speed.
6. Allow the weld to cool before handling or inspecting.

Expert Advice:

Common Mistake: Using incorrect amperage for the electrode and material thickness.

Solution: Consult welding charts for proper amperage settings.

Usage Tip: Practice on scrap metal to perfect technique before welding on important projects.

Demolition Hammer

A powerful tool designed for heavy-duty breaking and demolition tasks, such as breaking concrete, asphalt, and masonry.

Uses:

Breaking up concrete slabs, removing tiles, demolishing walls, and other heavy-duty demolition tasks.

Technical Details:

- Impact energy ranges from 10 to 60 joules.
- Speed ranges from 900 to 3000 blows per minute (BPM).
- Weight typically ranges from 5 to 30 kg.

Step-by-Step Guide:

1. Select the appropriate chisel bit for the task.
2. Insert the bit into the hammer.
3. Secure the work area and wear protective gear.
4. Hold the hammer firmly with both hands.
5. Start the tool and apply steady pressure.
6. Move the hammer to different areas as needed.

Expert Advice:

- Common Mistake: Applying too much pressure.
- Solution: Let the tool's weight and impact energy do the work.
- Usage Tip: Use a pointed chisel for precise breaking, and a flat chisel for larger surface areas.

Electric Oil Pump

A device used for transferring oil and other fluids efficiently and cleanly.

Uses:

- Transferring oil from drums, tanks and containers.

Technical Details:

- Flow rates typically range from 3 to 10 litres per minute (LPM).
- Power sources include electric and battery-operated.

Step-by-Step Guide:

1. Select the appropriate hose and nozzle.
2. Connect the pump to the power source.
3. Place the intake hose in the oil source.

4. Start the pump and monitor the flow.
5. Turn off the pump when the transfer is complete,

Expert Advice:

- Common Mistake: Overfilling containers.
- Solution: Monitor the flow rate and container capacity.
- Usage Tip: Go slow when the container is about to fill to prevent spills.

Concrete Vibrator

A tool used to eliminate air bubbles and ensure proper compaction of concrete.

Uses:

- Compacting concrete for foundations, walls, and slabs.

Technical Details:

- Vibration frequency typically ranges from 10,000 to 20,000 vibrations per minute (VPM).
- Shaft lengths range from 3 to 20 feet.

Step-by-Step Guide:

1. Insert the vibrator into the wet concrete.
2. Turn on the tool and slowly move it through the concrete.
3. Ensure complete coverage of the area.
4. Turn off the tool and remove it slowly and steadily.

Expert Advice:

- Common Mistake: Over-vibrating the concrete.
- Solution: Limit vibration time to prevent segregation.
- Usage Tip: Use in a systematic pattern for even compaction and do not miss a spot.

Paint Mixture

A tool used for mixing paint, coatings, and other viscous materials.

Uses:

- Mixing paint, adhesives, and cementitious materials.

Technical Details:

- Speed ranges from 500 to 900 RPM.
- Paddle parts include rod and spiral.

Step-by-Step Guide:

1. Attach the paddle to the mixer.
2. Place the mixer in the paint container.
3. Start the mixer and move it through the paint.
4. Mix until the paint reaches a uniform consistency,

Expert Advice:

- Common Mistake: Mixing at too high a speed.
- Solution: Use a lower speed to prevent air bubbles.
- Usage Tip: Clean the paddle immediately after use to prevent hardening.

Conclusion:

In this chapter, you have explored the diverse range of power tools categorised by their functions. Understanding the purpose and proper use of each tool is crucial for achieving efficiency, precision, and safety in your projects. Continue to the next chapter to learn practical techniques for using these tools effectively.

Chapter 3:

Mastering Common Techniques for Efficiency and Safety

Mastering the use of power tools requires more than just understanding their functions; it involves developing the skills and techniques to use them efficiently and safely. This chapter focuses on essential methods for achieving the best results with your power tools. Whether you are drilling, screwdriving, grinding, polishing, or working with precision tools like routers and planers, the techniques outlined here will help you enhance your efficiency and ensure safety. Additionally, maintaining sharp blades and bits is crucial for precision and performance, and this chapter will delve deeper into these topics.

This chapter will cover:

1. Essential drilling techniques.
2. Successful screwdriving.
3. Using grinders and polishers.
4. Achieving precision with tools like routers, planers, and moulding machines.
5. Efficiency and Precision: By Using Sharp Blades and Bits.

Learn how to maximise the potential of your power tools while prioritising safety and efficiency. Whether you're a professional or a DIY enthusiast, these techniques will help you achieve better results in your projects.

Essential Drilling Techniques Unveiled

Essential drilling techniques ensure accurate and efficient hole creation, while minimising the risk of damage to the workpiece or the drill itself. Here's an explanation of each technique:

1. Centre Punching to Mark the Drilling Point Accurately:

a. Before drilling, it's crucial to mark the centre point of the hole accurately. This prevents the drill bit from wandering and ensures the hole is drilled in the desired location.

b. A centre punch is a tool with a pointed tip used to make a small indentation or dimple at the drilling point. This dimple serves as a guide for the drill bit.

c. To centre punch, place the tip of the punch on the drilling point and strike it with a hammer. The indentation created will help keep the drill bit centred during drilling.

2. Maintaining Consistent Pressure Throughout the Drilling Process:

a. Consistent pressure ensures smooth and uniform drilling. Too much pressure can cause the drill bit to overheat or dull prematurely, while too little pressure may result in slower progress or the drill bit getting stuck.

b. Apply steady and firm pressure on the drill, allowing it to penetrate the material gradually. Avoid pushing too hard as this can strain the drill motor and lead to poor drilling results.

c. Pay attention to the sound and feel of the drill. Adjust the pressure as needed to maintain a steady drilling pace and prevent the drill from binding or seizing.

3. Ensuring the Drill Remains Perpendicular to the Surface:

a. Drilling at an angle can result in off-centre or misaligned holes, compromising the integrity of the workpiece.

b. Use a level or square to ensure the drill is positioned perpendicular (90 degrees) to the surface before starting the drilling process.

c. Maintain this perpendicular orientation throughout the drilling process by keeping the drill handle aligned with the surface and applying even pressure on the drill.

d. If drilling at an angle is intentional, ensure it is done accurately and consistently to achieve the desired outcome, without compromising the structural integrity of the material or workpiece.

Screw It! Mastering the Art of Perfect Fastening

Selecting the appropriate screwdriver bit, applying the right amount of torque, and using proper alignment techniques are essential for successful screwdriving.

1. Selecting the Appropriate Screwdriver Bit:

a. Match the screwdriver bit type to the screw head shape. Common types include Phillips, slotted (flathead), hex, Torx, and square (Robertson).

b. Ensure the bit size matches the screw size to prevent stripping or damaging the screw head.

c. Choose bits made from high-quality materials, such as hardened steel, for durability and longevity.

2. Applying the Right Amount of Torque:

a. Use a screwdriver with a handle that provides a comfortable grip and allows for adequate torque transmission.

b. Apply steady pressure while turning the screw to prevent slipping or stripping the screw head.

c. Avoid using excessive force as it can lead to overtightening, stripping the screw, or damaging the material.

3. Using Proper Alignment Techniques:

a. Position the screwdriver bit squarely into the screw head to ensure maximum contact and minimise the risk of slippage.

b. Hold the screwdriver handle perpendicular to the screw head and apply downward pressure to maintain contact.

c. Start screwing at a slow and controlled pace to establish proper alignment before increasing the speed.

4. Additional Tips:

a. For stubborn screws, consider using a penetrating lubricant or applying slight pressure while turning the screwdriver to break any corrosion or rust bonds.

b. If the screw head begins to strip, stop immediately to prevent further damage. Use a screw extractor, or an alternative screwdriving method, to remove the damaged screw.

c. Regularly inspect screwdriver bits for signs of wear or damage, and replace them as needed to maintain optimal performance.

Unleashing the Smooth Power of Grinders and Polishers

Using angle grinders, bench grinders, and polishers effectively requires attention to technique to achieve desired results and ensure safety.

1. Selecting the Right Abrasive:

a. Choose the appropriate abrasive wheel or attachment based on the material being worked on and the desired finish.

b. For angle grinders, select cutting discs for metal cutting, grinding discs for surface preparation, and wire brushes for rust removal.

c. Bench grinders typically use grinding wheels for sharpening tools and shaping metal.

d. Polishers may use polishing pads or buffing wheels with polishing compounds for achieving a smooth and glossy finish.

e. Ensure the abrasive is compatible with the grinder or polisher's spindle size and speed rating to prevent accidents.

2. Applying Even Pressure:

a. Maintain consistent and even pressure on the grinder or polisher to achieve uniform results and prevent excessive wear on the abrasive.

b. Avoid pressing too hard as this can cause the abrasive to wear out quickly, overheat the workpiece, or create gouges.

c. Use gentle yet firm pressure, allowing the abrasive to do the work without forcing it.

3. Maintaining Control:

a. Hold the grinder or polisher with both hands, keeping a firm grip on the handle(s) to maintain control during operation.

b. Stand in a stable position with feet shoulder-width apart to ensure balance and stability.

c. For angle grinders, maintain a secure grip on the auxiliary handle (if provided) to control the tool's direction and minimise kickback.

d. Keep the grinder or polisher moving smoothly and steadily across the workpiece, avoiding sudden movements or jerks that could result in uneven finishes or damage.

4. Additional Tips:

a. Always wear appropriate personal protective equipment (PPE), including safety glasses or goggles, hearing protection, and gloves, to protect against debris and noise.

b. Use the grinder or polisher at the correct speed and RPM (revolutions per minute) to avoid overheating and premature wear.

c. Regularly inspect abrasive discs, wheels, or pads for signs of wear, damage, or imbalance and replace them as needed to maintain safety and performance.

Precision Craftsmanship: Routers, Planers, and Moulding Machines

Achieving precise shapes and profiles with tools like routers, planers, and moulding machines requires attention to detail and proper technique. Here are some tips to help you:

1. Selecting the Right Bits and Blades:

a. For routers, choose router bits with the appropriate profile or cutting edge to achieve the desired shape or profile. Common types include straight bits, flush-trim bits, chamfer bits, and round-over bits.

b. Consider the material you'll be working with when selecting router bits and blades. Carbide-tipped bits are ideal for hardwoods and dense materials, while high-speed steel (HSS) bits are suitable for softer woods and plastics.

c. Planer blades should be sharp and properly aligned to ensure smooth and consistent cuts. Replace dull blades promptly to maintain optimal performance.

d. Moulding machines may use a variety of cutter heads or blades to create different profiles. Select the appropriate cutter head or blade based on the desired moulding profile and material.

2. Techniques for Precision:

a. Measure and mark the desired shape or profile on the workpiece before starting to ensure accuracy.

b. Secure the workpiece firmly using clamps or a vice to prevent movement during cutting or shaping.

c. Start with light passes and gradually increase the depth of cut to achieve the desired shape or profile, while minimising tear-out and splintering.

d. Maintain a consistent feed rate and keep the tool moving smoothly to avoid uneven cuts and chatter marks.

e. Use guide fences, templates, or jigs to help guide the tool and ensure uniformity across multiple workpieces.

3. Safety Precautions:

a. Always wear appropriate personal protective equipment (PPE), including safety glasses or goggles, hearing protection, and dust masks, to protect against debris and noise.

b. Keep hands and fingers away from moving parts and cutting edges while the tool is in operation.

c. Use push sticks or push blocks to safely feed material through planers and moulding machines, especially when working with small or narrow pieces.

4. Maintenance and Care:

a. Regularly clean and lubricate routers, planers, and moulding machines to ensure smooth operation and extend tool life.

b. Check for loose or worn parts, such as bearings, belts, and guide rails, and replace them as needed to maintain accuracy and safety.

c. Sharpen or replace router bits, planer blades, and moulding machine cutters as soon as they show signs of dullness or damage, to maintain clean and precise cuts.

Maximising Efficiency and Precision: Sharp Tools, Sharp Results

Maintaining sharp blades and bits is essential for both efficiency and safety. Regular sharpening or replacing of dull components ensures tools operate smoothly, accurately and safely, reducing the risk of accidents and enhancing the quality of work.

1. Efficiency Improvement:

- **a.** Cutting Speed: Sharp blades and bits cut through materials more quickly, reducing the time needed to complete tasks. This increased cutting speed translates to higher productivity.
- **b.** Energy Conservation: Efficient cutting reduces the energy consumption of the tool, whether it's powered by electricity, batteries, or manual effort. This can prolong the battery life of cordless tools and reduce power consumption in corded tools.

2. Precision Enhancement:

- **a.** Clean Cuts: Sharp edges make cleaner cuts with less effort, minimising rough edges and reducing the need for additional finishing work. This is particularly important in woodworking, where precision is crucial.
- **b.** Accuracy: Sharp blades and bits maintain their cutting path more effectively, resulting in more accurate cuts and holes. This precision is essential for tasks requiring exact measurements and detailed work, such as cabinetry and metal fabrication.

Dull Tools, Dangerous Work: The Hidden Risk of Blunt Tools.

1. Increased Strain:

a. On the Tool: Dull blades and bits require more force to cut through materials, putting additional strain on the motor and other components. This increased load can lead to overheating, faster wear, and potential breakdowns.

b. On the User: Users must exert more effort when using dull tools, leading to fatigue and potential strain injuries. Prolonged use of dull tools can cause repetitive strain injuries (RSIs) and increase the risk of accidents due to user fatigue.

2. Safety Hazards:

a. Slippage and Kickback: Dull tools are more likely to slip or bind in the material, causing sudden and unpredictable movements known as kickback. This can lead to serious injuries, especially when using high-powered tools like table saws and chainsaws.

b. Loss of Control: Increased effort required to push dull tools through materials can result in a loss of control, making it harder to maintain a steady hand. This loss of control can cause unintended cuts or holes and increase the risk of accidental injury.

c. Burn Marks: Dull blades generate more friction, which can cause burning or scorch marks on the material, particularly in woodworking. This not only damages the workpiece but also increases the risk of fire.

Tool Talk: When to Sharpen, or Preferably Replace

1. Decreased Performance:

a. Slow Cutting/Drilling: If a tool is cutting, drilling, or driving more slowly than usual, it likely needs sharpening.

b. Increased Effort: When more effort is required to use the tool, it indicates dullness.

2. Quality of Work:

a. Rough Edges: Cuts and holes have rough or jagged edges instead of clean, smooth finishes.

b. Tearing or Chipping: Wood or another material tears or chips instead of cutting smoothly.

3. Physical Inspection:

a. Visible Dullness: The cutting edge appears rounded or worn.

b. Nicks and Chips: The blade or bit has visible nicks, chips, or cracks.

4. Noise and Vibration:

a. Unusual Sounds: The tool makes unusual noises during operation, which can indicate increased friction.

b. Excessive Vibration: Increased vibration during use can signify a dull blade or bit.

5. Heat Generation:

a. Overheating: The tool generates more heat than usual, indicating it is working harder than necessary due to dullness.

By following these guidelines, you can ensure your tools remain in optimal condition, providing efficient and precise performance. Regular maintenance, and timely sharpening or replacement, will extend the life of your tools and improve the quality of your work.

General Check for Sharpness

1. Based on Usage Frequency:

a. Daily Use: For tools used daily, such as those in professional workshops, check every week.

b. Weekly Use: For tools used a few times a week, check every month.

c. Occasional Use: For tools used occasionally, check every 3-6 months.

2. Based on Material Type:

a. Softwoods: Tools used on softwoods may require less frequent sharpening as the material is less abrasive.

b. Hardwoods: Tools used on hardwoods should be sharpened more frequently due to the increased wear and tear.

c. Metal and Masonry: Tools used on metal or masonry should be sharpened or checked more regularly, potentially after every major project due to the high wear and tear.

Conclusion:

By mastering these essential techniques, you will not only improve the quality of your work, but also ensure your safety while using power tools. Each method discussed in this chapter contributes to achieving efficiency, precision, and professional results in your projects. Continue to apply these practices and refine your skills as you progress, making the most of your power tools.

Chapter 4:

Tool Care Mastery: Maintenance and Safety

SAFETY FIRST

Power tools are essential for efficiency and precision that manual tools can't match. However, their power and speed also present significant safety risks if not used properly. This comprehensive guide provides crucial safety guidelines for a wide range of power tools, from cutting and drilling equipment to woodworking and gardening tools. Prioritising safety when using power tools is important. Let's explore the key principles and specific recommendations for safe power tool operation.

General Safety Guidelines for Power Tools

1. Read the manual: Always thoroughly read and understand the manufacturer's instructions before using any power tool.
2. Wear appropriate Personal Protective Equipment (PPE),
3. Inspect tools before use: Check for damaged parts, loose fittings or frayed cords.
4. Maintain a clean work area: Keep your workspace organised and free of clutter to prevent accidents.

5. Avoid distractions: Stay focused on the task at hand and avoid using power tools when tired, or under the influence of drugs or alcohol.
6. Use the right tool for the job: Don't force a tool to do a job it wasn't designed for.
7. Keep tools sharp and clean: Properly maintained tools are safer and more efficient.
8. Secure your work: Use clamps or a vise to hold your work when practical.
9. Disconnect tools when not in use: Unplug tools before changing accessories or making adjustments.
10. Store tools safely: Keep tools in a dry, secure place, out of reach of children.
11. Never carry a tool by the cord. This can damage the cord contact and create a safety hazard.
12. Keep children away: Ensure that children and bystanders are at a safe distance from the work area.

Tool-Specific Safety Guidelines

Cutting Tools (Angle Grinder, Circular Saw, Etc.)

- Always use the guard that comes with the tool.
- Ensure blades or discs are appropriate for the tool and material being cut.
- Allow the tool to reach full speed before contacting the work surface.
- Never force the cut; let the tool do the work.
- Be aware of kickback, and plan your cut accordingly.

Drilling & Driving Tools (Drill Machine, Impact Drill, Etc.)

- Secure loose clothing and tie back long hair to prevent entanglement.
- Use the appropriate bit for the material being drilled.
- Apply steady pressure when drilling, but don't force the tool.
- Be prepared for the drill to bind, especially when breaking through materials.

Grinding & Polishing Tools (Angle Grinder, Car Polisher, Etc.)

- Allow the tool to come to a complete stop before setting it down.
- Never use a grinding wheel that has been dropped or is cracked.
- Don't grind or cut with the side of a disc unless it's designed for that purpose.
- Use both hands to maintain control of the tool.

Woodworking Tools (Router, Planer, Jigsaw, Etc.)

- Always work against the cutter rotation or blade direction.
- Keep cutting tools sharp to prevent binding and kickback.
- Secure small workpieces in a vice or with clamps.
- Allow tools like planers and routers to reach full speed before contacting the work surface.

Cleaning Tools (Pressure Washer, Vacuum Cleaner, Etc.)

- Never point a pressure washer at people or animals.
- Be cautious of electrical components when using wet vacuums.
- Wear appropriate PPE to protect against debris and water spray.

Garden Tools (Chainsaw, Brush Cutter, Etc.)

- Be especially aware of your surroundings and potential hazards.
- Maintain proper footing and balance at all times.

  For gas-powered tools, refuel in a well-ventilated area and when the engine is cool.
- Keep chainsaw chains sharp and properly tensioned.

Miscellaneous Tools (Air Compressor, Paint Sprayer, Etc.)

- Regularly drain moisture from air compressor tanks.

  When using paint sprayers, ensure proper ventilation and wear a respirator.

  For welding machines, use appropriate eye protection and fire-resistant clothing.

- With concrete vibrators, be aware of electrical hazards, especially in wet conditions.

Remember, these guidelines are general and should be used in conjunction with the specific instructions provided by the manufacturer for each tool. Always prioritise safety and use common sense when operating power tools.

Personal Protective Equipment (PPE)

By understanding the importance of PPE and recognising common hazards associated with power tool use, individuals can take proactive measures to protect themselves and minimise the risk of injuries in the workplace.

1. Importance of Personal Protective Equipment (PPE) for Safety:
    a. A PPE serves as a vital barrier between the user and potential hazards, minimising the risk of injury.
    b. Proper PPE can significantly reduce the severity of injuries in case of accidents.
2. Common Hazards and Risks:
    a. Flying Debris: Power tools, especially cutting and grinding tools, can generate flying debris that may cause eye injuries or lacerations.
    b. Loud Noise: Prolonged exposure to loud noise from power tools can lead to hearing loss or damage.
    c. Exposure to Harmful Substances: Certain power tools, such as those involving paints or chemicals, can expose users to harmful fumes or substances, leading to respiratory issues or skin irritation.
3. Types of Personal Protective Equipment (PPE):
    a. Eye Protection: Wear safety glasses or goggles to shield the eyes from flying debris.
    b. Hearing Protection: Use earplugs or earmuffs to reduce the risk of hearing damage from loud noise.
    c. Respiratory Protection: Use respirators or dust masks when working with materials that produce dust, fumes, or harmful particles.

d. Hand Protection: Wear gloves to protect hands from cuts, abrasions or chemical exposure.

e. Body Protection: Wear appropriate clothing, such as long sleeves and pants, to protect the skin from cuts, burns or chemical splashes.

f. Foot Protection: Wear steel-toed boots or shoes with reinforced toes to prevent foot injuries from heavy objects or falling debris.

Compliance and Enforcement (India):

The Factories Act, 1948 and the Occupational Safety, Health and Working Conditions Code, 2020.

Under these Acts, it is the responsibility of the employer to ensure that appropriate PPE is provided to employees working with machinery or construction tools to minimise the risk of accidents and occupational hazards. Failure to comply with these provisions can result in legal consequences.

Emergency Procedures:

a. Maintain a list of emergency contacts (e.g. local emergency services, poison control).

b. Keep a well-stocked first aid kit in the workspace.

c. Fire Safety: Use fire extinguishers and ensure safe storage of flammable materials.

What to Do in Case of an Accident:

a. Stopping the tool immediately.

b. Providing first aid for minor injuries.

c. Seeking medical attention for serious injuries.

Essential Safety Checklist:

- ☐ Wear appropriate PPE.
- ☐ Inspect tools before use.
- ☐ Ensure proper ventilation.
- ☐ Keep the work area clean and organised.
- ☐ Follow the manufacturer's instructions.
- ☐ Disconnect the power source when not in use.

Maintenance Magic: Keeping Your Tools in Top Shape

Power tools are invaluable assets in any workshop, construction site, or project. However, to ensure these tools continue to perform at their best and remain safe to use, regular maintenance is crucial.

Proper maintenance of power tools encompasses a range of practices, including regular cleaning, proper storage, routine inspections, timely repairs, and adherence to manufacturer guidelines. These practices not only extend the life of your tools, but also enhance their performance, ensure safety during use, and can even save you money in the long run by preventing costly replacements.

Let us explore the essential aspects of power tool maintenance, from basic cleaning techniques to advanced care procedures. We'll also delve into proper storage methods, routine inspection protocols, and critical safety guidelines. By following these maintenance practices, you can keep your power tools in top shape, ready to tackle any job with efficiency and safety.

Remember, a well-maintained tool is a reliable tool. Let's dive into the world of power tool maintenance and discover how to keep your equipment running smoothly for years to come.

Preventative Care: Secrets to Tool Longevity

1. Regular Cleaning: Keep tools clean to prevent the build-up of dust and debris.
2. Proper Storage: Store tools in a dry, cool place to prevent moisture damage and corrosion.
3. Routine Inspection: Regularly check tools for wear and tear and address issues promptly.

Quick Maintenance Checklist:

- ☐ Clean tools after each use.
- ☐ Inspect for wear and damage.
- ☐ Lubricate moving parts as needed.
- ☐ Check and tighten loose components.
- ☐ Store tools properly.

The Art of CLEANING

Regular cleaning of power tools is crucial for their longevity and performance because accumulated dust, debris, and residues can significantly impact their operation. Over time, these contaminants can infiltrate the tool's internal components, causing friction, overheating, and mechanical blockages. This build-up can lead to wear and tear on moving parts, reducing the tool's efficiency and lifespan. Additionally, dust and debris can clog ventilation systems, leading to overheating and potential motor damage. Residues from materials such as adhesives or paint can also affect the smooth operation of parts. By keeping power tools clean, users can ensure smoother operation, reduce the risk of malfunction, and extend the overall durability and reliability of their equipment.

Your Go-To Cleaning Products:

1. Brushes (soft and stiff),
2. Compressed air.
3. Blower.
4. Cleaning cloths (microfiber or lint-free).
5. Mild cleaning solution (mild soap and water, or specific tool cleaning solutions).
6. Lubricant (machine oil or grease).
7. Vacuum cleaner (with attachments for small areas).

The Ultimate Tool Cleaning Guide:

1. Unplug and Disassemble:

a. Power Supply Disconnection: Ensure the tool is unplugged, or the battery is removed.

b. Part Removal: Disassemble removable parts, such as blades, bits, or sanding pads. For example, remove the blade from a saw or the bit from a drill.

Note: These are the most important safety steps before cleaning and maintenance. It avoids any kind of accident and electrical shock.

2. Brush Cleaning:

a. Exterior Surfaces: Use a stiff brush to remove dirt and debris from the exterior housing of the tool.

b. Delicate Parts: Use a soft brush to clean more delicate or intricate parts, such as vents, guards, or chucks.

3. Compressed Air: Internal Cleaning:

Blow compressed air into the ventilation slots and other hard-to-reach areas to remove internal dust and debris.

Example: Use compressed air to clean the motor housing of an angle grinder, or the battery compartment of a cordless drill.

4. Cleaning Solution:

a. Surface Wiping: Dampen a cloth with a mild cleaning solution and wipe down the exterior surfaces of the tool.

   Example: Clean the handle and body of a power sander, or the casing of a circular saw.

b. Part Cleaning: Clean removable parts, such as blades, discs, or belts, with a suitable cleaning solution.

5. Reassemble and Lubricate:

a. Reassembly: Once all parts are dry, reassemble the tool.

   Example: Reattach the blade to a saw, or the bit to a drill.

b. Lubrication: Apply a small amount of lubricant to moving parts such as bearings, spindles, or pivot points.

Example: Lubricate the chuck of a drill, or the moving parts of a jigsaw.

Safety Tips:

- Always wear safety gear, including gloves and safety glasses, when cleaning power tools.
- Follow the manufacturer's instructions for cleaning and maintenance.
- Ensure tools are completely dry before reassembly and use.
- Store tools in a clean, dry place to prevent dust accumulation and corrosion.

STORAGE

Importance of Proper Storage

1. Extending Tool Life:

    a. Protection from Physical Damage: Proper storage prevents power tools from being accidentally dropped, bumped, or scratched, reducing wear and tear.

    b. Avoiding Misplacement: Storing tools in designated places ensures they are easy to find, reducing the risk of them being lost or misplaced.

    c. Humidity Control: Humidity can cause rust and corrosion on metal parts. Proper storage in a dry environment can prevent this.

    d. Dust and Dirt Protection: Keeping tools in cabinets, toolboxes, or using covers protects them from dust and dirt, which can clog moving parts and affect performance.

2. Safety:

    a. Preventing Accidents: Proper storage reduces the risk of accidents, such as tripping over tools or being injured by sharp edges.

    b. Child Safety: Storing tools out of reach of children prevents accidental usage and potential injuries.

3. Organisational Efficiency:

    a. Improved Workflow: Having a well-organised storage system helps in quickly locating and accessing tools, enhancing productivity.

    b. Inventory Management: Proper storage allows for easy inventory checks, ensuring that tools are available and in good condition when needed.

By ensuring proper storage and considering environmental factors, you can significantly extend the life of your power tools and maintain their performance and safety.

Tips for Organising Tools

Categorise and group similar tools:

1. By Type: Organise tools by their function, such as drilling tools, cutting tools, measuring tools, etc.
2. By Frequency of Use: Place frequently used tools in easily accessible locations and store less frequently used tools in higher or lower spots.
3. By Project Type: Group tools commonly used together for specific projects or tasks.

Storage Options

By choosing the right storage solution for each type of tool, you can enhance organisation, improve accessibility and ensure the longevity of your power tools.

Toolboxes:

- Description: Portable containers designed to store and organise tools.
- Types: Available in various sizes, from small handheld boxes to large rolling chests.

Recommended For: Ideal for small hand tools, screwdrivers, wrenches, pliers and drill bits.

- Benefits: Portability, easy access, and protection from dust and debris.

Cabinets:

- Description: Stationary storage units with multiple shelves and drawers.

- Types: Range from small wall-mounted units to large, heavy-duty floor cabinets.

  Recommended For: Suitable for larger power tools, such as drills, saws, grinders, and sanders.

- Benefits: Provides secure, organised storage with ample space for larger tools and accessories.

Pegboards:

- Description: Wall-mounted boards with holes designed to hold hooks and tool holders.
- Types: Metal and wooden pegboards, with various hook configurations.

  Recommended For: Perfect for small hand tools, screwdrivers, pliers, hammers and frequently used items.

- Benefits: Easy access, customisable layout and maximises vertical space.

- Shelves:

- Description: Flat, horizontal surfaces for storing tools and supplies.
- Types: Available as standalone units or wall-mounted systems.

  Recommended For: Suitable for medium to large tools, toolboxes, and storage bins.

- Benefits: Versatile, provides clear visibility of stored items and is easy to install.

Tool Carts:

- Description: Mobile storage units with multiple drawers and shelves.

- Types: Available in different sizes and configurations, often with locking wheels.

  Recommended For: Ideal for workshops and garages where tools need to be moved around frequently.

- Benefits: Portability, organised storage, and convenient access to tools at different workstations.

Drawers and Drawer Organisers:

- Description: Compartments within cabinets or tool chests for detailed organisation.
- Types: Can be built-in or add-on organisers for existing drawers.

  Recommended For: Great for small tools, parts, fasteners, and accessories.

- Benefits: Keeps small items sorted, prevents clutter and makes it easy to locate specific tools or parts.

Wall-Mounted Racks:

- Description: Racks installed on walls to hold tools vertically.
- Types: Can be designed for specific tools, such as screwdrivers, hammers, and clamps.
- Recommended For: Best for frequently used tools and for keeping workspaces clear.
- Benefits: Saves floor space, provides quick access, and helps keep tools organised and visible.

Heavy-Duty Cabinets:

- Description: Robust, secure storage units designed to hold large and heavy tools.

- Types: Metal cabinets with reinforced shelves and locking mechanisms.
- Recommended For: Suitable for storing large power tools, such as table saws, miter saws and drill presses.
- Benefits: Secure storage, protects valuable tools and supports heavy loads.

INSPECTION

Routine Inspection of Power Tools

Regular inspection of your power tools is crucial for maintaining safety and ensuring optimal performance. Follow these steps to conduct a thorough routine inspection:

1. **Visual Check:**
   - Examine the tool's body for cracks, dents, or other damage.
   - Inspect cords and plugs for fraying, exposed wires, or bent prongs.
   - Check for loose, missing or damaged parts.
2. **Guard and Safety Feature Inspection:**
   - Ensure all guards are in place and functioning correctly.
   - Test safety switches and triggers for proper operation.
   - Verify that blade or bit guards retract and return as designed.
3. **Moving Parts Assessment:**
   - Check for proper alignment of moving parts.
   - Look for signs of excessive wear or damage on blades, bits, or other attachments.
   - Ensure all parts move freely without binding or excessive play.
4. **Electrical Component Examination:**
   - Inspect brushes in electric motors for wear, if accessible.
   - Check battery connections and housing for corrosion or damage (for cordless tools).
   - Verify that all electrical connections are secure.

5. **Lubrication Check:**
    - Ensure moving parts are properly lubricated, as per the manufacturer's instructions.
    - Look for signs of oil leaks or insufficient lubrication.
6. **Cleanliness:**
    - Remove dust, debris and build-up from all parts of the tool.
    - Clean vents and air intakes to prevent overheating.
7. **Functional Test:**
    - Operate the tool briefly (without load) to check for unusual noises, vibrations, or smells.
    - Verify that the tool reaches full speed and maintains it consistently.
8. **Accessory Inspection:**
    - Check the condition of blades, bits, and other accessories.
    - Ensure they are sharp, undamaged and appropriate for the tool.
9. **Storage Condition Assessment:**
    - Verify that tools are stored in a clean, dry place.
    - Ensure they are protected from extreme temperatures and humidity.
10. **Documentation Review:**
    - Keep maintenance records up to date.
    - Check if any tools are due for professional servicing or calibration.

Perform this inspection before each use and conduct a more thorough check periodically, depending on the frequency of use. If any issues are found, address them immediately, either through repair or replacement, before using the tool. Remember, a well-maintained tool is a safer tool.

Inspection Calendar Based on Usage Frequency:

a. Daily Use: For tools used daily, such as those in professional workshops, inspect every week.

b. Weekly Use: For tools used a few times a week, inspect every month.

c. Occasional Use: For tools used occasionally, inspect every 3-6 months.

Professional Power Tool Inspection Checklist

Tool Name: ___________________, Date: __________

Inspector:

Model/Serial Number:

1. General Condition.
   - ☐ Tool body free of cracks, dents or damage.
   - ☐ All parts are present and securely attached.
   - ☐ Labels and warnings legible.
2. Electrical Components.
   - ☐ Power cord intact, no fraying, or exposed wires.
   - ☐ Plug undamaged, prongs straight.
   - ☐ Battery (if applicable) free of damage, holds charge.
   - ☐ On/off switch functioning properly.
3. Guards and Safety Features.
   - ☐ All guards present and properly attached.
   - ☐ Guards move freely and return to their positions.
   - ☐ Safety switches/interlocks operational.
4. Moving Parts
   - ☐ Parts aligned correctly.
   - ☐ No excessive wear or damage.
   - ☐ Smooth operation, no binding.
5. Cutting/Working Surfaces.
   - ☐ Blades/bits should be sharp and undamaged.
   - ☐ Correct type for the tool.
   - ☐ Securely fastened.

6. Lubrication
   - □ Moving parts adequately lubricated.
   - □ No signs of oil leaks.
7. Cleanliness
   - □ Tool, free of dust and debris.
   - □ Vents and air intakes clear.
8. Functional Test
   - □ Tool starts and stops correctly.
   - □ Reaches and maintains full speed.
   - □ No unusual noises, vibrations or smells.
9. Accessories
   - □ All necessary accessories are present.
   - □ Accessories in good condition.
10. Storage
    - □ Tool stored in a clean, dry place.
    - □ Protected from extreme conditions.

Additional Notes:

____

____

____

Action Required: □ None, □ Repair, □ Replace, □ Further Inspection

Inspector Signature: __________________, Date: ____________

Next Inspection Due: ____________________

This checklist provides a comprehensive overview of the key areas to inspect for most power tools. It can be adapted or expanded based on the specific type of tool being inspected. Regular use of this checklist can help ensure that tools are maintained in a safe working condition and that potential issues are identified and addressed promptly.

Conclusion

Proper maintenance and adherence to safety protocols are essential for the longevity and effective use of your power tools. This chapter has provided you with comprehensive guidelines for cleaning, storage, maintenance, and safety practices. By following these recommendations, you can ensure that your tools remain in excellent condition, ready to deliver optimal performance for any task. Remember, a well-maintained tool not only enhances efficiency but also significantly reduces the risk of accidents, ensuring a safe and productive working environment. Carry these practices forward to maintain the reliability and safety of your power tools for years to come.

Chapter 5:

Power Tools Across Industries

In various trades, power tools play an essential role in enhancing efficiency, precision, and safety. This chapter will explore the use of power tools across different professions, detailing the specific tools, their applications, and examples of projects where they are indispensable.

Construction and Demolition

Demolition Hammer

- Use: Breaking down concrete, brick, and other hard materials.
- Applications: Demolition of old structures, roadwork, and renovation projects.
- Projects: Tearing down walls for a home renovation, breaking up old driveways and dismantling concrete foundations.

Rotary Hammer

- Use: Drilling and chiselling hard materials, such as concrete and masonry.

  Applications: Construction of buildings, installation of heavy-duty anchors.

- Projects: Installing anchor bolts for structural steel, drilling holes for rebar, and chiselling out channels for plumbing.

Core Drill

- Use: Drilling large holes in concrete, stone, and asphalt.
- Applications: HVAC installations, plumbing and electrical work.
- Projects: Drilling holes for HVAC ductwork, creating openings for plumbing pipes, and installing electrical conduits through concrete.

Wall Chaser

- Use: Cutting grooves into walls for laying electrical wiring or pipes.
- Applications: Electrical installations, plumbing and construction.

Projects: Cutting channels for electrical wiring in new constructions, creating grooves for plumbing pipes in bathroom renovations.

Automotive and Detailing

Car Polisher

- Use: Polishing and buffing vehicle surfaces.

  Applications: Automotive detailing, maintenance.

  Projects: Buffing out scratches on car bodies, applying wax for a shiny finish, and restoring headlight clarity.

Pressure Washer

- Use: High-pressure cleaning of surfaces.
- Applications: Vehicle cleaning, driveway cleaning, and outdoor surfaces.
- Projects: Washing mud off cars, cleaning oil stains from driveways, and removing dirt from outdoor patios.

Washer Gun

- Use: Attachment for pressure washers, for detailed cleaning.

  Applications: Automotive detailing, spot treatment.

- Projects: Cleaning intricate areas of car rims, targeting stubborn spots on concrete, and washing tight spaces on outdoor furniture.

Woodworking and Carpentry

Circular Saw

- Use: Making straight cuts in wood, plywood and other materials.

  Applications: Framing, cabinetry, and general carpentry.
- Projects: Building wooden decks, cutting plywood for furniture, and framing house walls.

Miter Saw

- Use: Making precise crosscuts and miters in wood.
- Applications: Trim work, framing and furniture making.
- Projects: Cutting crown moulding for interior decoration, framing doorways, and windows, and constructing picture frames.

Electric Router

- Use: Shaping edges, creating grooves and decorative work.

  Applications: Cabinetry, joinery, and decorative woodworking.
- Projects: Creating decorative edges on tabletops, routing grooves for cabinet doors, and making intricate joinery for furniture.

Planer

- Use: Smoothing and thinning wood boards.
- Applications: Furniture making, door fitting, and general carpentry.

  Projects: Smoothing rough-sawn lumber, thinning boards for custom furniture, and fitting wooden doors to frames.

Jig Saw

- Use: Making curved and intricate cuts in wood.
- Applications: Custom woodworking, furniture making, and crafts.
- Projects: Cutting out shapes for wooden signs, making custom shelves with curved edges, and crafting intricate wooden puzzles.

Electrical and Plumbing

Drill Machine

- Use: Creating holes and driving screws.

  Applications: Electrical installations, plumbing, and general construction.

- Projects: Drilling holes for electrical outlets, installing screws for plumbing fixtures, and assembling framing for walls.

Impact Drill

- Use: Drilling into hard materials, such as concrete and masonry.
- Applications: Electrical work, plumbing installations, construction.

  Projects: Mounting electrical panels on concrete walls, drilling anchor holes for plumbing pipes, and installing heavy-duty fixtures.

Cordless Screwdriver, Electric Screwdriver

- Use: Inserting screws and fasteners.

  Applications: Electrical installations, assembling fixtures, general maintenance.

- Projects: Securing electrical boxes, assembling furniture, and installing cabinet hardware.

Metalworking

Angle Grinder

- Use: Cutting, grinding and polishing metal.
- Applications: Metal fabrication, automotive repair and welding.
- Projects: Cutting metal rods for custom frames, grinding welds on metal structures, and polishing metal surfaces for a smooth finish.

Bench Grinder

- Use: Sharpening tools, grinding metal and polishing.
- Applications: Tool maintenance, metal preparation, and finishing.
- Projects: Sharpening drill bits and chisels, grinding down metal burrs, and polishing metal components for machinery.

Cut Off Machine

- Use: Making straight, precise cuts in metal.

  Applications: Metal fabrication, construction, and manufacturing.
- Projects: Cutting steel pipes for plumbing, trimming metal sheets for roofing, and slicing metal bars for structural work.

Welding Machine

- Use: Joining metal pieces through welding.
- Applications: Construction, manufacturing, and automotive repair.
- Projects: Welding steel beams for building construction, repairing metal components on vehicles, and fabricating custom metal furniture.

Landscaping and Agriculture

Backpack Brush Cutter, 4-Stroke

- Use: Cutting through thick vegetation.
- Applications: Landscaping, garden maintenance, and clearing land.
- Projects: Clearing overgrown fields, trimming dense bushes around properties and maintaining large garden areas.

Brush Cutter, 2-Stroke / 4-Stroke

- Use: Trimming grass and clearing brush.
- Applications: Garden maintenance, landscaping.
- Projects: Trimming grass along fences, clearing undergrowth in gardens, and maintaining paths in parks.

Earth Auger

- Use: Drilling holes in the ground.
- Applications: Planting trees, installing fences and landscaping.
- Projects: Drilling holes for fence posts, preparing soil for tree planting and installing deck supports.

Home Improvement and Maintenance

Wet & Dry Vacuum Cleaner.

- Use: Cleaning both wet and dry debris.
- Applications: Construction sites, workshops, and home improvement.
- Projects: Cleaning sawdust in woodworking shops, removing spilled liquids in garages, and clearing debris from renovation sites.

Heat Gun

- Use: Applying heat to various materials.
- Applications: Paint stripping, bending plastics, and drying materials.
- Projects: Stripping old paint from doors, bending PVC pipes for plumbing, and drying plaster before painting.

Electric Pump

- Use: Moving fluids.
- Applications: Water transfer, irrigation, and dewatering.

Projects: Pumping water from flooded basements, transferring water for garden irrigation, and dewatering construction sites.

Painting and Finishing

Paint Mixer

- Use: Mixing paint and other liquids.

  Applications: Painting and coating work.
- Projects: Mixing paint for large painting projects, blending plaster for wall finishes, and preparing sealants for application.

Electric Vibrator

- Use: Consolidating concrete.
- Applications: Construction, ensuring proper settling of concrete.
- Projects: Vibrating concrete for foundation pours, consolidating concrete for driveway installations, and ensuring smooth finishes for concrete walls.

Electric Sander (Orbital, Belt, or Electric)

- Use: Smoothing surfaces and removing finishes.
- Applications: Furniture restoration, home improvement projects, woodworking.
- Projects: Sanding old paint off furniture, smoothing wooden floors before refinishing, and preparing walls for repainting.

Air Compressor

- Use: Powering pneumatic tools and inflating tyres.
- Applications: Workshops, automotive repair, construction.

  Projects: Running pneumatic nail guns, inflating vehicle tyres, and operating spray paint guns.

Stone and Masonry

Marble Cutter

- Use: Cutting marble and other stone materials.

  Applications: Tile work, stone masonry, countertop fabrication.

- Projects: Cutting marble tiles for bathroom installations, shaping stone for kitchen countertops, and trimming stone pavers for outdoor walkways.

Slab Cutter

- Use: Cutting large stone slabs.
- Applications: Masonry, stone fabrication, construction.
- Projects: Slicing granite slabs for monuments, cutting large stone for building facades, and shaping marble slabs for flooring.

General Maintenance and Utility

Air Blower

- Use: Blowing away dust, debris and drying surfaces.
- Applications: Cleaning, maintenance, drying.
- Projects: Clearing leaves from driveways, drying wet surfaces after cleaning and blowing dust off workbenches.

By understanding the specific applications and benefits of these power tools, professionals across various trades can enhance their efficiency, precision and safety, leading to better results and more successful projects.

Breaking Boundaries: Innovative Use of Power Tools

Power tools have revolutionised various industries, from construction and woodworking to metalworking and artistic projects. While most users are familiar with standard applications, this chapter delves into innovative and advanced uses that can significantly enhance efficiency and creativity. We'll explore these applications with a strong emphasis on safety, practical tips, and sustainability.

Before each section, you'll find a skill level indicator: Beginner, Intermediate, Advanced

Remember, regardless of skill level, always prioritise safety. Wear appropriate protective gear, including safety goggles, gloves, ear protection, and dust masks. Follow manufacturer guidelines and maintain your tools regularly.

Angle Grinder

Metal Sculptures: Shape and sculpt metal using various grinding discs to create unique art pieces.

Safety: Wear a full-face shield, heat-resistant gloves, and fire-resistant clothing. Work in a well-ventilated area, away from flammable materials.

Tips: Begin with thinner metals and progress to thicker ones. Use different disc grits for rough shaping, and fine detailing.

Lawn Mower Blade Sharpening: Sharpen lawn mower blades using a grinding disc.

Safety: Wear gloves and goggles. Secure the blade firmly in a vice.

Tips: Maintain the original blade angle. Sharpen both sides evenly, and check for balance after sharpening.

Circular Saw

● Drywall Cutting: Make precise cuts in drywall for custom wall openings using a fine-toothed blade.

Safety: Wear a dust mask and safety goggles. Be aware of potential electrical wiring or plumbing behind the drywall.

Tips: Score the cutting line with a utility knife first. Use a drywall saw for smaller, more intricate cuts.

Core Drill

● Planting Holes: Create perfect planting holes in hard soil or clay for landscaping projects.

Safety: Wear safety goggles and steel-toed boots. Be aware of underground utilities.

Tips: Start with a smaller pilot hole. Add water to the hole to make drilling easier in hard soil.

● Door Peephole Installation: Drill precise holes for peephole installation in doors.

Safety: Wear safety goggles. Secure the door to prevent movement during drilling.

Tips: Measure and mark both sides of the door to ensure alignment. Use a drill stop to prevent drilling too far.

● Furniture Cable Management: Create neat holes for cable management in desks or entertainment centres.

Safety: Use a drill guide for precise holes. Wear safety goggles.

Tips: Measure and mark hole locations carefully. Protect surfaces from scratches. Use appropriate bit sizes.

Impact Drill

● Removing Rusted Bolts: The hammering action can help loosen rusted or stuck bolts.

Safety: Wear safety goggles and gloves. Be prepared for sudden movement, if the bolt breaks free.

Tips: Apply penetrating oil and let it sit before attempting removal. Use the correct size bit to avoid stripping the bolt.

Rotary Hammer

- Tile Removal: Use the hammering action and chisel attachment for efficient tile removal during renovations.

Safety: Wear safety goggles, ear protection, and a dust mask. Be cautious of flying debris.

Tips: Start at a grout line or loose tile. Work in small sections to maintain control.

Cordless Screwdriver.

- Wind Up Extension Cords: Use a winding attachment to efficiently wind up extension cords or holiday lights.

Safety: Keep fingers away from the winding mechanism.

Tips: Guide the cord evenly as it winds. Use low speed to avoid tangling.

- Open Tight Jar Lids: Use a rubber attachment to open tight jar lids with ease.

Safety: Ensure the jar is stable and use a low speed setting.

Tips: Apply gentle, consistent pressure. If the lid doesn't budge, try alternating between clockwise and counterclockwise motions.

Orbital Sander

- Remove Paint: Remove old paint from intricate woodwork using coarse-grit sandpaper.

Safety: Wear a dust mask, safety goggles and gloves. Test for lead paint in older homes.

Tips: Start with coarse-grit and finish with finer grits. Use a vacuum attachment to minimise dust.

Faux Finish Preparation: Prepare surfaces for faux painting techniques using medium-grit sandpaper.

Safety: Wear a dust mask and safety goggles, and ensure proper ventilation.

Tips: Sand in circular motions for an even surface. Clean thoroughly before applying paint.

Heat Gun

Remove Stickers: Use a low heat setting to remove stubborn stickers or decals, without damaging the surface.

Safety: Work in a well-ventilated area. Be cautious of fumes from adhesive residue.

Tips: Start with low heat and increase gradually if needed. Use a plastic scraper to assist removal.

Culinary Uses: Various food preparation techniques.

Safety: Use food-safe attachments. Keep heat away from flammable materials.

Tips:

- Melting Cheese: Move the heat gun constantly to prevent burning.
- Roasting Marshmallows: Use low heat settings for even browning.
- Coffee Roasting: Stir beans constantly for even roasting.

- Reheating Barbecue: Use moderate temperatures to prevent drying out.

Advantages: Precision temperature control, smoke-free operation, and no chemical odours.

Paint Sprayer

🟠 Create Ombre Effects: Use multiple sprayers for colour gradients on furniture, blending one colour into another.

Safety: Wear a respirator mask and full-body protection. Ensure proper ventilation.

Tips: Start with the lightest colour and gradually introduce darker shades. Use a smooth, sweeping motion.

🟡 Fabric Dyeing: Apply even, consistent dye to large fabric pieces.

Safety: Use fabric-specific dyes. Wear a respirator mask and protective clothing.

Tips: Test on a small area first. Protect surrounding areas from overspray. Apply in thin, even coats.

Chainsaw

🟠 Ice Fishing: Use a clean, oil-free chain to cut holes in thick ice for ice fishing.

Safety: Wear ice cleats, a life jacket, and bring ice safety picks. Never work alone on ice.

Tips: Use a guide to ensure straight cuts. Clean and dry the chain thoroughly after use to prevent rust.

🟠 Ice Sculpting: Create intricate ice sculptures for events or competitions.

Safety: Use a clean, oil-free chain. Work in a cold environment. Wear thermal gloves, and non-slip footwear.

Tips: Secure the ice block firmly. Work from a stable position. Use gentle, controlled movements.

- Tree Sculpting: Create artistic tree sculptures or topiary designs.

Safety: Use a carving bar and chain. Secure the trunk firmly before carving. Wear protective gear.

Tips: Plan your design beforehand. Work from a stable position. Start with larger shapes, and refine details.

- Shaping Hedges: Create precise, artistic shapes in hedges and shrubs.

Safety: Wear safety goggles and gloves, and ensure stable footing.

Tips: Start with basic shapes and gradually add complexity. Step back frequently to assess the overall shape.

Pressure Washer

- Rug/Mat Cleaning: Use low pressure settings to clean outdoor rugs and mats.

Safety: Wear waterproof boots and eye protection. Be cautious of slippery surfaces.

Tips: Test a small, inconspicuous area first. Use an appropriate cleaning solution for the material.

Air Blower

- Drying Surfaces: Use a narrow nozzle for quickly drying surfaces after washing, such as cars or outdoor furniture.

Safety: Wear safety goggles. Be cautious of flying debris.

Tips: Start from the top and work your way down. Use sweeping motions for even drying.

- Paint Texturing: Create unique textured effects on painted surfaces.

Safety: Use low air pressure. Wear eye protection, and a respirator mask.

Tips: Experiment with distance and angle for desired effects. Practice on scrap materials first.

- Inflating Air Mattresses: Quickly inflate air mattresses and other inflatable items.

Safety: Use appropriate nozzle attachments, and don't overinflate.

Tips: Check for leaks before fully inflating. Use low pressure for delicate items.

Conclusion

Exploring these advanced applications of power tools can significantly enhance your productivity and creativity across various projects. Whether you're a professional contractor, a DIY enthusiast, or an artist, understanding these innovative uses opens up new possibilities. Always prioritise safety by wearing appropriate protective gear and following manufacturer guidelines to ensure safe and effective use of power tools.

Remember, mastering these advanced techniques takes practice. Start with simpler projects and gradually work your way up to more complex applications. With patience and persistence, you'll be able to tackle a wide range of innovative projects using your power tools.

Chapter 6:

The Rise of Cordless Power Tools: Unleashing Freedom

Evolution of Battery Technology

The evolution of battery technology has significantly transformed cordless power tools, starting with early developments powered by heavy and inefficient nickel-cadmium (NiCd) batteries, which were known for their durability but also for their weight and memory effect that reduced charge capacity over time. The transition to Nickel-Metal Hydride (NiMH) batteries marked an improvement, offering higher energy density and longer run times, although still limited by weight and charge duration. The revolution came with Lithium-Ion (Li-ion) batteries, which provided a significant leap forward with higher energy density, longer run times, and lighter weight. Innovations like rapid charging technology reduced downtime during projects, while smart battery management systems increased safety and efficiency by preventing overcharging, overheating, and deep discharge.

Why Go Cordless?

- **Portability:**
  - Cordless tools eliminate the need for power cords, providing unparalleled freedom of movement

and making them ideal for remote or confined workspaces without power outlets.

- **Ease of Use:**
    - Without cords, setup is simpler and faster. Users avoid the hassle of managing tangled cords or locating extension cables.
- **Versatility:**
    - The market offers a wide range of cordless tools, from drills and saws to sanders and grinders, catering to various applications in construction, woodworking, metalworking, and home maintenance.

Understanding Cordless Limitations:

- **Power Output:**
    - While cordless tools have advanced, they generally offer lower power output compared to their corded counterparts. For heavy-duty tasks, corded tools might still be necessary.
- **Battery Life:**
    - Despite improvements, battery run time can still be a concern. For prolonged tasks, users may need spare batteries or frequent recharging.
- **Cost:**
    - Cordless tools and their batteries tend to be more expensive initially than corded tools. Additionally, the long-term cost of replacing batteries can add up.

Top Picks: Most Popular Cordless Tools

- **Cordless Drill:**
  - Use: Drilling holes and driving screws into various materials, such as wood, metal, and plastic.
  - Applications: Construction, furniture assembly, home repairs, and DIY projects.
  - Advantages: Portable, versatile, essential for many tasks around the house, and on job sites.
- **Cordless Hammer:**
  - Use: Hammering and drilling into tough materials, such as concrete and masonry.
  - Applications: Construction, demolition, and heavy-duty repair work.
  - Advantages: Powerful impact force, portability, essential for tasks requiring both drilling and hammering actions.
- **Cordless Blower:**
  - Use: Clearing debris, dust and leaves from work areas.
  - Applications: Yard maintenance, workshop cleaning, and construction site cleanup.
  - Advantages: Lightweight, easy to manoeuvre, ideal for quick and efficient clean-ups without cords.
- **Cordless Chainsaw:**
  - Use: Cutting through wood and tree branches.
  - Applications: Tree trimming, logging, landscaping, and emergency storm cleanup.

    - Advantages: Mobility, reduced noise compared to gas-powered saws, and suitable for various cutting tasks.

- **Cordless Angle Grinder:**
    - Use: Grinding, cutting and polishing metal and masonry.
    - Applications: Metalworking, masonry work, automotive repair, and construction.
    - Advantages: Freedom of movement, capable of handling a variety of tasks, including cutting, grinding, and polishing.

- **Cordless Wrench:**
    - Use: Tightening and loosening nuts and bolts.
    - Applications: Automotive repair, machinery maintenance, and construction.
    - Advantages: High torque, portability, ideal for tasks requiring quick and efficient fastening.

- **Cordless Screwdriver:**
    - Use: Driving screws and small fasteners.
    - Applications: Furniture assembly, home repairs, and light construction work.
    - Advantages: Compact design, ease of use, perfect for repetitive screwdriving tasks.

Cordless power tools have revolutionised the way many trades operate, providing unparalleled convenience and efficiency. As battery technology continues to advance, the capabilities of these tools will only expand, making them indispensable in both professional and DIY settings.

Cordless Care Checklist:

- ☐ Proper battery charging and storage.
- ☐ Maintaining clean battery contacts,
- ☐ Rotating battery use for even wear.
- ☐ Storing tools without batteries for long-term storage.
- ☐ Disposing of old batteries properly.

Chapter 7:

Smart Tools: The Future of Power Tools

The future of power tools is heading towards smarter, more connected and highly efficient devices. As technology advances, power tools are not just becoming more powerful and portable, but also more intelligent, transforming the way we work and maintain these essential tools.

Futuristic Technologies

Integration of Digital Displays:

- Real-time Feedback: Future power tools are expected to come equipped with digital displays that provide real-time feedback on tool settings, battery life, and performance metrics. This information can help users make informed decisions on the go, improving efficiency and ensuring optimal tool performance.
- User Interface: These displays will likely feature intuitive interfaces, making it easier for users to adjust settings, monitor tool health, and access diagnostic information without needing to refer to manuals or external devices.

Smart Sensors:

- Material Detection: Tools will increasingly incorporate sensors that can detect and adjust to different materials automatically. This capability will optimise performance by adjusting speed, torque, and power output based on the material being worked on, reducing wear and tear, and enhancing precision.

Condition Monitoring: Sensors will also monitor tool conditions, such as temperature, vibration, and load. This will enable predictive maintenance, alerting users before a tool fails and thereby reducing downtime and repair costs.

Platforms and Apps:

- Tool Management: Connected platforms and apps will allow users to manage and control their tools remotely. These platforms can track tool usage, schedule maintenance, and provide updates and alerts about tool status.
- Enhanced Productivity: Users will be able to optimise tool usage through data analytics provided by these platforms, leading to better project planning and execution. Apps could also offer tutorials, troubleshooting guides, and customer support to enhance the user experience.

Internet of Things (IoT) Enhancements:

- Safety Features: IoT integration will bring enhanced safety features to power tools. Tools will have automatic shutdown capabilities in case of overheating, overloading, or other malfunctions. Geofencing can prevent unauthorised use by restricting tool operation to specific locations, while emergency alerts can notify users of potential hazards.

- Remote Monitoring: Users will be able to monitor tool status and receive alerts in real-time, ensuring that tools are always in optimal condition and used safely.

Futuristic Tools Already in Use:

- ### Robotic Lawn Mowers:

One of the most exciting developments in power tools is the emergence of autonomous tools. Robotic lawn mowers, for instance, can operate independently, maintaining lawns with minimal human intervention. These machines use sensors and GPS to navigate and cut grass efficiently.

- ### Robotic Vacuum Cleaners:

Similar to robotic lawn mowers, robotic vacuum cleaners are already in use, autonomously cleaning floors in residential and commercial spaces. They use sensors to navigate and avoid obstacles, ensuring thorough cleaning with minimal human intervention.

Automated Cutting Machines:

The future may also see the rise of automated cutting machines that can perform precise cuts without direct human control. Such tools could be programmed for specific tasks, enhancing productivity in manufacturing and construction environments.

The integration of smart technologies into power tools marks a significant shift towards greater efficiency, safety, and convenience. As these advancements continue to evolve, the tools we use will become increasingly adept at adapting to our needs, leading to more streamlined and effective work processes in various industries. The future of power tools is undoubtedly smart, paving the way for innovations that will revolutionise how we approach our tasks and projects.

Appendix: Your Power Tool Companion

Tool Selection Guide

Choosing the right tools for your projects is crucial for achieving optimal results. With a multitude of options available in the market, selecting the best tools can often be overwhelming. This guide aims to simplify the process by providing key criteria to consider when evaluating and comparing different tool brands and models.

- **Quality and Durability:**

Assessing the build quality and durability of tools is essential for ensuring longevity and reliability. Look for brands known for using high-quality materials and robust construction techniques that can withstand the rigours of regular use in demanding environments.

- **Performance:**
    - Evaluate the performance of tools based on factors such as power output, efficiency, and user feedback. Consider metrics like speed, torque, cutting precision, and overall effectiveness in completing tasks. Reviews from professionals and DIY enthusiasts can provide valuable insights into real-world performance.

- **Innovation:**

Consider how each brand incorporates new technologies and features into their tools. Look for innovations that enhance productivity, safety, and user experience. Features like smart sensors, ergonomic designs, and advanced battery technology can significantly improve tool performance and usability.

- **Warranty and Support:**
    - Warranty terms and customer support services are crucial aspects to consider when choosing tools. Evaluate the duration and coverage of warranties offered by different brands, including provisions for repairs, replacements, and customer service responsiveness. A reliable warranty and responsive support can provide peace of mind and assurance of long-term investment value.

- **Value for Money:**
    - Price is undeniably a critical consideration, but it›s imperative to weigh it against the tool›s overall value. Evaluate features, performance, and durability relative to the price to ascertain the best value within your budget. While it may be tempting to opt for cheaper alternatives, resist compromising on quality or essential features. Investing in quality tools upfront can lead to long-term savings by enhancing productivity and minimising maintenance costs over time. Therefore, prioritise value over short-term savings to ensure optimal performance and longevity from your investment.

By carefully considering these criteria and conducting thorough research, you can confidently select tools that meet your specific requirements and deliver exceptional performance and reliability.

Troubleshooting Common Issues

Power tools are indispensable in various trades, offering efficiency and precision. However, like any machinery, they can encounter issues that hinder their performance. This chapter provides a comprehensive guide to troubleshooting and solving common problems faced while using power tools. Understanding these issues and their solutions can save time, enhance productivity and extend the lifespan of your tools.

1. Blade Jams

Symptoms: The tool stops abruptly, the blade does not move, or there is excessive noise.

Causes: Cutting through dense material, dull blades, or debris build-up.

Solutions:

- Check for Debris: Turn off the tool and unplug it. Inspect the blade, and remove any lodged material.
- Sharpen or Replace Blades: Ensure the blade is sharp, and replace it if necessary.
- Use Appropriate Blade: Make sure you are using the correct blade for the material being cut.
- Avoid Overloading: Do not force the tool; let the blade do the cutting.

2. Motor Issues

Symptoms: Tool fails to start, motor overheats, or there is a burning smell.

Causes: Overheating, electrical issues or worn-out components.

Solutions:

- Inspect Power Supply: Check the power source and connections. Ensure the tool is plugged in properly, or the battery is charged.
- Allow Cooling Time: If the motor overheats, turn off the tool and let it cool down.
- Check Brushes: Inspect the motor brushes for wear, and replace if necessary.
- Clean Vents: Ensure the air vents are clean and unobstructed to allow proper cooling.

3. Battery Malfunctions

Symptoms: Battery does not charge, short run time, or inconsistent power.

Causes: Battery degradation, poor connections, or charger issues.

Solutions:

- Check Connections: Ensure the battery is properly seated, and connections are clean.
- Test Charger: Use a known good charger to rule out charger issues.
- Battery Care: Follow proper battery care practices, such as not overcharging, and storing in a cool, dry place.
- Replace Battery: If the battery is old or damaged, consider replacing it with a new one.

4. Excessive Vibration

Symptoms: Unusual shaking or vibration during operation.

Causes: Imbalanced attachments, loose parts, or worn bearings.

Solutions:

- Tighten Attachments: Ensure all blades, bits, and attachments are securely fastened.

  Inspect Bearings: Check for worn or damaged bearings, and replace if needed.

- Balance Attachments: Use balanced and compatible attachments to reduce vibration.

5. Power Loss

Symptoms: Tool loses power during use or operates intermittently.

Causes: Electrical issues, faulty switches, or damaged cords.

Solutions:

Inspect Cords: Check for cuts, frays, or other damage to power cords, and replace if necessary.

- Test Switches: Ensure the power switch is functioning correctly, and replace if faulty.
- Check Battery: Ensure the battery is fully charged and functioning properly.

6. Overheating

Symptoms: Tool becomes excessively hot during use.

Causes: Continuous use without breaks, poor ventilation, or internal friction.

Solutions:

- Allow Cool Down Periods: Use the tool in intervals to prevent overheating.

- Clean Tool Regularly: Ensure that vents and internal components are free from dust and debris.
- Lubricate Moving Parts: Apply appropriate lubricants to reduce friction and heat.

Glossary: Your Power Tool Lingo

1. Arbour: The shaft or spindle on which cutting blades or attachments are mounted.
2. Arbour Nut: The nut is used to secure cutting blades or attachments onto the arbour of a power tool.
3. Auxiliary Handle: An additional handle or grip on a power tool to provide better control and stability during operation.
4. Blade Guard: A safety feature that covers the cutting blade when not in use, to prevent accidental contact.
5. Chuck: The mechanism used to hold drill bits or other attachments in place on a power tool.
6. Chuck Capacity: The maximum diameter of drill bits or attachments that a chuck can accommodate.
7. Chuck Key: A specialised tool used to tighten or loosen the chuck on certain power tools.
8. Depth Stop: A feature that allows users to set a specific drilling or cutting depth to achieve uniform results.
9. Kickback: The sudden, forceful backward movement of a power tool caused by the binding or snagging of the blade or bit.
10. Overload Protection: A safety feature that automatically shuts off the tool if it detects excessive heat or current draw.
11. RPM: Revolutions Per Minute, indicating the speed at which a tool's motor rotates.
12. Torque: The rotational force generated by a power tool, measured in foot-pounds (ft-lb) or Newton-metres (Nm).

13. Trigger: The switch or button that controls the operation of a power tool.

14. Trigger Lock: A safety feature that prevents accidental activation of the tool by locking the trigger in the off position.

15. Variable Speed: The ability to adjust the operating speed of a power tool to suit different materials and tasks.

About the Authors

Kunal Garg

With 18 years of experience in the power tools business in India, Kunal Garg has established himself as a proficient analyst and manager. His extensive knowledge of the industry, combined with his hands-on experience, has enabled him to navigate the complexities of the power tools market effectively. Kunal's expertise in business strategy and market analysis continues to drive the success and growth of the family business.

Mayank Garg

Mayank Garg also brings 18 years of experience in the power tools business, with a specialised focus on purchasing and manufacturing tools from China factories. His expertise is rooted in his deep understanding of quality control and manufacturing processes, gained from living in Yongkang, China—a city known as the world's largest supplier of power tools. Mayank's in-depth research and collaboration with local manufacturers have been instrumental in ensuring the high-quality and reliability of the tools produced under his supervision.

www.ingramcontent.com/pod-product-compliance
Ingram Content Group UK Ltd.
Pitfield, Milton Keynes, MK11 3LW, UK
UKHW060359300726
14090UKWH00001B/22

* 9 7 9 8 8 9 5 1 9 8 3 8 4 *